Invisible Rainbows

Invisible Rainbows

The Unseen Universe Beyond Our Senses

Dr Alfredo Carpineti

First published in 2026 in the UK by Wilton Square Books

29 Wilton Square
London N1 3DW
www.wiltonsquarebooks.com

A CIP record for this book is available from the British Library

1 3 5 7 9 8 6 4 2

ISBN 978-1-80677-044-1 (hardback)
ISBN 978-1-80677-045-8 (ebook)

Typeset by Phil Beresford

Printed in Great Britain by CPI Group UK Ltd, Croydon CR0 4YY

MIX Paper | Supporting responsible forestry FSC® C013604

Our authorised representative in the EU for product safety is: Logos Europe, 9 rue Nicolas Poussin, 17000, La Rochelle, France
contact@logoseurope.eu

For all the queer kids reaching for the stars.

CONTENTS

INTRODUCTION
The Colours of the Unknown

Me vero primum dulces ante omnia Musae,
quarum sacra fero ingenti percussus amore,
accipiant caelique vias et sidera monstrent,
defectus solis uarios lunaeque labores.

Me first before all things may the sweet Urania,
Whose rites I bear with mighty passion pierced,
Receive, and show the paths and stars of heaven,
The sun's eclipses and the labouring moon.

—*Virgil*

When I feel down about the state of the world, about the indifference of the Universe, I always think of rainbows. Rainbows are a trick of the light that appear when light encounters rain. The colours that make up what we consider white light are made visible, as water droplets don't scatter all the colours in the same way. And suddenly – there it is: a rainbow, traditionally described as seven stripes, indigo, violet, blue, green, yellow, orange and red. Sometimes you get a double rainbow. If you are high enough you can see it go all the way round in a circle, with no ends for leprechauns to place gold. But there is another thing that makes rainbows special and it makes my little nerdy heart leap. Nobody else sees the same rainbows as you do, even if they are standing right next to you. Light doesn't work that way. It moves in straight lines from the rain droplets to your eyes, and to your eyes alone. The rainbows that you see are just for you. No one else in the entire Universe ever has or ever will see exactly that rainbow. I love a rainbow – and not just because it's the colours of the flag of my people; it is a pretty thing that hides a profound truth about ourselves. That our senses don't tell us everything. You see, a rainbow, even in its uniqueness, tells us something about the way we understand the Universe. It shows us that our senses are limited. We see sunlight as colourless. Where have these colours been hiding? They're right there; it's just that our eyes deceive us. Are there even more colours that we do not see? The answer is a resounding *yes*.

But the story of this book starts elsewhere. A few years ago I was at a science festival, making cocktails onstage while talking about astronomy, when I was asked if rainbows exist on other worlds. Obviously we could speculate on planets around distant stars, but we are pretty confident that other rainbows, though not quite as we know them, exist in our own galaxy. There is a moon of Saturn called Titan. Titan is the biggest satellite of Saturn, bigger than the planet Mercury. It is also the only other place in the solar system with rivers, lakes, seas and, most importantly, the only moon with rain. But it's not the rain we are familiar with on Earth. It is far too cold on Titan for that. The liquid there is methane (and other hydrocarbons) and it rains occasionally, although not as often as on Earth. If you were in a spaceship around Saturn and you looked at Titan, you would see a hazy orange world: nothing too exciting – you wouldn't see the lakes and rivers. But if you could see Titan in infrared, the light that is emitted by warm bodies, the clouds would suddenly become transparent and you would catch glimpses of sunlight reflecting on those lakes. Spacecraft that we have sent there have seen them. But let me take you a step further. Imagine yourself on the cold surface of Titan. It is raining and the Sun is shining through some clouds. There in the sky, using infrared, you would see a rainbow, very similar to Earth's own but visible only to you. It's an invisible rainbow to everyone else, but it still exists. As unique and incredible as the one that we can experience here.

I wanted to explore the wonder of these invisible rainbows, and more, in this book. Scientists divide light into seven broad categories: radio waves, microwaves, infrared, visible, ultraviolet, X-rays and gamma rays. Radio waves have the longest wavelength, and we have been using them for telecommunications. Gamma rays are the shortest and the most energetic. I will explore each invisible light to show how astronomers are using them to discover new wonders in the new Universe. In this book, you will hear first-hand from the people actually making these remarkable discoveries. They are there at the final frontier, unlocking the mysteries of the cosmos. They are, to me, the most exciting astronomers of our time.

Scientific work does not exist in a vacuum; its context in the social and historical sphere where it sits, matters. For example, we cannot discuss infrared light and ignore the controversy surrounding the naming of the largest space telescope after James Webb, who was involved in the mass dismissal of homosexual people from the US government in the 1950s. This is one of many examples, showing that, while science is incredibly precious to humanity, it can also be messy and political. We have to talk about the human element. Today, there is so much to learn beyond the visible that it's almost restrictive trying to fit it all in one book.

I had to make a selection. Not to be too Italian about it and bring it all down to food, but *Invisible Rainbows* will act as a tasting menu for what the Universe beyond visible light is like, and your complimentary cocktail will be served in the chapter about radio waves. Let's take this extraordinary journey together. And it really is extraordinary. To experience what we're about to learn, we would need the ability to see beyond our senses. First, we'll be faced with what's invisible in the things that we cannot see, before finding the invisible even in what is bright and right in front of us. But first, let there be visible light.

What Is Light?

Light is a spectrum, specifically the electromagnetic spectrum. The cosmos is awash with a sea of electric and magnetic fields, and we call that light. For a long time, we called light a tiny fraction of it all: the part that those of us privileged enough to see can perceive. This is visible light, or the visible spectrum. But since the 1800s we've known there is a lot more: infrared, ultraviolet, radio and microwaves, X-rays and gamma rays. But culturally we do struggle with this a bit. Radio waves still intrinsically feel like something that we *listen to* rather than observe. X-rays and gamma rays, with their links to radiation, are more often seen as dangerous particle physics. No matter the wavelength, light is fundamentally all the same. If unifying the long-wavelength radio waves with the short wavelength gamma rays is already a struggle, I'm afraid we will have to take a tumble down the rabbit hole. I won't be asking you to believe six impossible things before breakfast. Just one more. We've got to accept that light is two things at once. Light is waves that travel across vacuum and transparent materials, and light is also particles, which we call photons, delivering energy to matter in discrete packets. Every type of light is *both* a photon and a wave. For example, we call radio waves *radio waves* but we can equally call them *radio wave photons*. The duality of light is a fact, but one so contradictory that it took thousands of years to get there. The greatest minds of humanity, from Euclid to Ibn Sina, from Dignāga to Newton, argued that light was made of (either) particles or waves. They were all half right, but before quantum mechanics they could not imagine that properties seen as opposite might coexist in the same object. Reality is a lot weirder than we think it is.

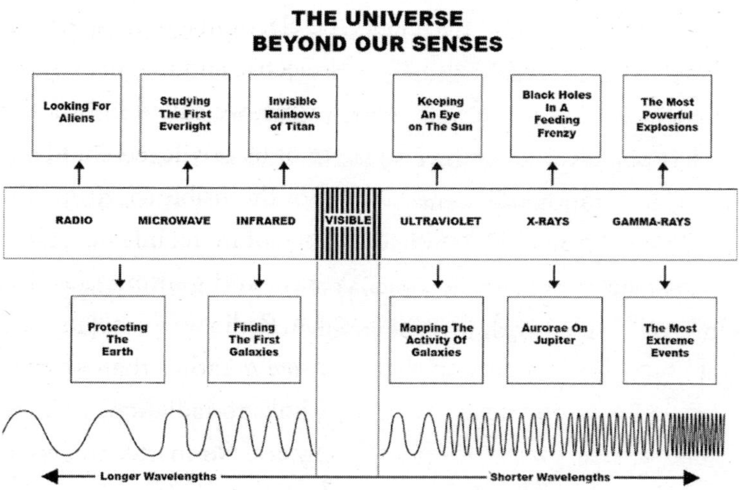

The electromagnetic spectrum and a few astronomical uses of the various wavelengths of light.

Alfredo & Chris Carpineti

In our experience, the whole world is a merry-go-round of colours, shades, shimmer and shine based on the smallest fraction of the electromagnetic spectrum. For over 225 years we have known that there is so much that we cannot see. For less than half that time we have discovered incredible astronomical features, thanks to our technological ability to venture beyond what is visible to us. We have begun to see the Universe not how we expected it to be, but as it actually *is*.

Thanks to light beyond the visible, we will learn about the most powerful explosions in the Universe and the treasures they produce. Light was not always free to move in the Universe and it was not always produced by its inhabitants, so thanks to the invisible light we will see the first, second and third time that light shone in the Universe. Some of this light will let us discover distant galaxies and even explore the connection between the birthplace of stars and the smell of raspberries (and, yes, this is related to the cocktail!). The deeper layers of the Sun will become apparent with this invisible light, as well; we will go from the hottest objects in the Universe to the coldest. Enormous black holes, and impossibly dense spinning neutron stars. Sure, the invisible rainbows of Titan have beckoned us already, but there are so many more new phenomena out there. Certainly, the entire Universe is filled with invisible rainbows.

— CHAPTER 1 —

RADIO WAVES

The Name of the Game Is Versatility

> Since we cannot change reality,
> let us change the eyes which see reality.
>
> —*Nikos Kazantzakis*

> Were all stars to disappear or die,
> I should learn to look at an empty sky
> And feel its total dark sublime,
> Though this might take me a little time.
>
> —*W. H. Auden*

Let us start with a recipe. We are making a cosmic cocktail with raspberry and rum. In a shaker, over ice, mix one part rum, one part raspberry syrup and one part lime juice. Shake and serve over ice in a tumbler. A mocktail alternative is almond milk and that raspberry syrup. Both are uncommon pairings, but trust me when I tell you how well they go together. The reason for that is in the molecular structure. The taste of raspberry and part of the smell of rum is caused by ethyl formate, an intriguing little molecule (you don't have to remember this name; there won't be a quiz at the end). What we are looking for is the right vibe. Focus on the idea of the cocktail, or make it – it's always five o'clock somewhere, thanks to the relativity of time.

Now, cocktail in hand, cast your mind 26,000 light years away: it's a huge, incomprehensible distance for our little ape brain. That's how far the centre of the Milky Way is. And there, among many incredible objects, including a supermassive black hole, there's a molecular cloud – a big collection of hydrogen peppered with other molecules. Among them is ethyl formate. A cloud at the centre of our galaxy that *almost* tastes of raspberry and smells of rum. (I did say almost.) The cloud is mostly hydrogen and the most common molecule that has some flavour is a cyanide … hence the almond milk for the mocktail version. Still, it is fun to imagine a deliciously deadly cloud out there.

We know of these molecules thanks to radio waves. Elements and molecules can absorb or emit different types of light, and astronomers use those ranges of the spectrum to discover them. Across all the wavelengths, we'll discover the building blocks of the Universe.

Giant satellite dishes popularised in movies and documentaries are often portrayed as our 'ear' to the sky. We are ever hopeful of listening to possible signals from extraterrestrials. We do listen to the radio. But radio waves are just another form of light, the one with the longest possible wavelengths. Radio waves are incredibly versatile when it comes to the study of the cosmos, a versatility that we will experience extensively in the topics in this chapter. But the origin of radio astronomy comes from the way it was used by our species at the beginning of the last century to communicate wirelessly. Antennae were built far and wide to bridge the distance that separated countries and people, and from this revolutionary development came the discovery of something else. Something incredible.

It is the early 1930s and our protagonist is Karl Jansky. He was working at Bell Labs in New Jersey studying the radio emission in the atmosphere. He discovered that the emission came mostly from nearby or distant thunderstorms. But there was also a 'hiss', a faint static of unknown origin which repeated every twenty-four hours. Well, not twenty-four hours exactly; every 23 hours and 56 minutes, to be precise. The devil is in the detail, as the saying goes. The Sun shines directly above each specific place on Earth every twenty-four hours. That's a *solar day*: the one we use. However, a star will reappear over the same position every 23 hours and 56 minutes, since the Earth has moved around the Sun a little over that day. This is known as a *sidereal day*.

And so, Jansky realised that the signal was coming to us from space, specifically from the direction of the constellation of Sagittarius, from the very centre of the Milky Way. After the detection was announced, the American public was lucky enough to hear the hiss broadcast on 15 May 1933 – with Jansky explaining that it came from well beyond the solar system. It was explained that the source was millions and millions of times more powerful than any radio station on Earth. Across the Atlantic, just days apart, something else was going on. The biggest book burning of the Nazi regime took place in Berlin in what is now known as Bebelplatz. It is estimated that 20,000 unique volumes of studies on intersex, gay and trans people were burned after a raid on Magnus Hirschfeld's pioneering Institut für Sexualwissenschaft (Institute for Sexual Science). Many thousands more volumes were burned that night, things that were considered un-German and immoral. That included volumes on Einstein's general relativity.

Einstein's work was referred to disparagingly by the Nazi regime's scientists as 'Jewish science'. Still, Einstein's theory was correct. Relativity itself was the key to solving the mystery of the hiss. The culprit turned out to be a supermassive black hole weighing 4.3 million times the mass of the Sun, sitting placidly at the centre of the Milky Way. In contrast to the persecution mentioned above, an international collaboration was formed to unravel this particular cosmic mystery, bringing us to the Event Horizon Telescope.

A Telescope as Big as the Earth

There is a wonderful property of light that is exquisitely exploited in radio astronomy. It's called interferometry and it allows us to make telescopes act as if they are much bigger than they actually are.

If you have two radio dishes a certain distance apart studying the same object at the same time, you can combine their observations. The result would be as if you were seeing that object with a telescope the size of the distance between the two. This is why you often see radio antennae in groups: you get more out of them. But our supermassive black hole, which we call Sagittarius A* (pronounced A-star), is very far away. It would be like seeing a doughnut or a bagel on the surface of the Moon. So you need a big telescope, one as big as the surface of the Earth. Radio telescopes from Greenland to the South Pole, from Europe to South America, and later, with the addition of Africa and Hawaii, looked towards the hiss and finally saw its shadow.[1] The doughnut or bagel analogy is a fitting one. The shadow is a bright emission surrounding a dark core, so very doughnut-like. For Sagittarius A*, the whole doughnut fits well within the orbit of Mercury, which has a diameter of 116 million kilometres. The image might look out of focus, but it allowed for the testing of some of the predictions of the theory of general relativity. All thanks to this international cooperation, which is just one of many in science. Working together has allowed humanity to transcend not just its limits, but even the limits of the instruments we have built to overcome them.

Human ingenuity is awesome, but there is the obvious underlying question: what the heck are we looking at when we look at Sagittarius A*? If you think black holes are by definition black, you'd be absolutely correct. Nothing escapes a black hole, not even light, and that is the fastest thing that exists in the known Universe. Basically, a black hole is an object whose gravitational pull is so strong that its escape velocity is greater even than light. With this in mind, it's easier to put a boundary around the black hole, which we call the event horizon. Once you cross the event horizon, you would have to move faster than the speed of light to be able to escape. As far as we know, that's impossible, so what goes into the black hole doesn't come out. These objects are fascinating and remain utterly mysterious because, at their very hearts, our physics fails. General relativity and quantum mechanics are pinnacles of human genius, but they do not work together in extreme events such as black holes or the Big Bang. Despite these limitations, scientists are exploring new theories, set to revolutionise once again our understanding of the Universe, its past and even its future.

But theories require observations both as a springboard to new insight and as a testbed for their prediction. And given that we can't really make a black hole in our backyard (or at CERN, as some conspiracy theorists believe), our best option is to look at black holes in the heavens. I sat down with Dr Sthabile Kolwa (she/they) to learn more about that. Dr Kolwa is a lecturer at the University of Johannesburg, studying galaxies and supermassive black holes with cutting-edge observatories such as MeerKAT – an array of sixty-four radio dishes located in South Africa. Dr Kolwa is brilliant – we have met multiple times in person and digitally – and each time I am in awe of their confidence in speaking, whether they are talking about science or what's good to eat in an Indian restaurant in Cape Town. Dr Kolwa explained:

> The emission that comes from black holes is detected over different wavelengths or energies of photons, spanning all the way from X-rays down to the radio. But the radio [emission] is particularly important because this is light that is not impeded or attenuated by dust. You've got dusty grains in a galaxy. Optical light will be blocked. We won't see it. Ultraviolet light will be blocked. But the radio waves will happily go through the dusty regions.

Their work with MeerKAT is cutting-edge, and the observatory, which is, unsurprisingly, the pride and joy of South Africa, has provided incredible results. Among the many results, it detected the extremely faint signals that provide the best view of the central region[2] of our galaxy yet – the few thousand light-years at the heart of the Milky Way. Right in the middle, surrounded by fast-moving stars, swirling gas clouds and peculiar magnetic structures, sits Sagittarius A* in all its majesty. Not bright enough to be studied outside radio wavelengths – we will discover throughout this book that the effects of a black hole's activities can be seen but not the black hole itself. And yet, even the quintessentially unseeable object cannot escape its relationship to the wider Universe. To misquote the poet and cleric John Donne, no celestial body is an island.

The Event Horizon Telescope looked at many objects, but two targets were different. One was our very own Sagittarius A*; the other was the supermassive black hole at the centre of elliptical galaxy M87 (known now as M87*). For these two targets, the virtual observatory sought the shadow of a black hole. If the name alone invokes some sci-fi movie, the explanation for it is even further beyond our experience. Einstein's work showed that *mass curves space-time*. The larger the mass, the larger the curvature. Dense objects can warp space-time so much that they act like magnifying glasses. The light of background objects can become magnified, allowing humanity to see even further into the Universe than possible with our instruments alone. Stars, planets, galaxies and clusters of galaxies are able to create this effect. And, of course, black holes as well. So rays of light that by chance are passing near one of these celestial plugholes are deflected by the extreme gravity of the black hole. A ray of light going straight to the black hole would be sucked in, but if it were to pass just shy of it, it would experience an incredible warp. This warped light creates the so-called photon ring, which depends strongly on how heavy the black hole is and how fast it is spinning. This photon ring is the 'shadow' cast by the black hole and is seen so beautifully in the case of M87* and Sagittarius A*.

If the only result from the entire field of radio astronomy was a single glimpse of the shadow of a supermassive black hole, all the work, effort and investment would have been worth it. But radio astronomy has versatility and advantages we do not see in other wavelengths. The achievement of the Event Horizon Telescope is certainly epochal, but radio waves also allow us to study planets and galaxies, dangerous asteroids, and search for intelligent life beyond Earth. Extreme objects and whole cosmic regions are accessible thanks to radio dishes spread out across the surface of our planet, on every continent, making it a truly global endeavour.

'Radio astronomy is particularly important because when we consider the electromagnetic spectrum, there are only two main observational windows that Earth's atmosphere provides us with,' Dr Kolwa readily answered my question on the importance of this discipline. They continued:

> Very high energy radiation, such as gamma rays and X-rays, is not able to get through the atmosphere, which is great for us. We're alive because of that. And also some infrared wavelengths are blocked. That means observing specific parts of the electromagnetic spectrum requires us to go to space. But for radio wavelengths, we don't have to go to space because the Earth's atmosphere is entirely transparent. There is a limitation in terms of time as well. Even though the Earth's atmosphere is transparent to optical wavelengths, you cannot observe during the day. Radio telescopes allow you to observe at any time of the day, unlike optical telescopes.

Astronomers believe that at the centre of almost every galaxy there is a supermassive black hole, just like Sagittarius A* sits at the centre of the Milky Way. There are different types of black holes, so there's a specific definition of supermassive. That's a black hole weighing between 100,000 times the mass of our Sun and tens of billions of times. M87* weighs 6.5 billion times our little yellow star. If you are wondering why we say 'almost every galaxy' – it's a get-out-of-jail-free card for our limitations. There is still plenty we do not understand, but as far we have found, each galaxy has its own supermassive black hole, and each one influences the others enormously. Galaxies are more than a simple collection of stars. They are changing, evolving organisms, equal parts familiar objects of cosmology and majestic mystery. In fact, calling them organisms is a shorthand – a little sci-fi extension of James Lovelock and Lynn Margulis's Gaia hypothesis – the idea that life on our own planet is so linked that the whole Earth can be considered an organism. It is not a completely unsound analogy. Galaxies *breathe in* intergalactic gas, and *exhale* jets and winds of plasma due to the actions of their supermassive black holes. They eat smaller galaxies, growing and changing. The influx of pristine material from the small galaxy absorbed in a galaxy merger revitalises the bigger one; new stars are born, and the supermassive black hole receives fresh food to feed on. This is enough to generate those jets and winds: black holes are messy eaters, spewing part of their meals back into intergalactic distances where they came from. Nothing escapes the event horizon of a black hole but, as you approach it, you might end up propelled on a trajectory sending you far beyond and into intergalactic space. These galaxies with active supermassive black holes are a favourite focus of so many observatories on Earth and in space. You want to be where the action is. And to those catalogues of data, radio astronomy is adding so much more. It sees what the other wavelengths cannot.

'The fields that we are now pointing the MeerKAT telescope at already have plenty of multi-wavelength detections. The new and sensitive radio detections we have will be a complement. And when we combine all of those together, we'll be able to obtain very powerful data sets that allow us to study galaxies and galaxy clusters in exceptional detail,' Dr Kolwa continued.

One thing to love about astronomers is how they talk about future observations. They are always in need of a bit more time, a bit more observation and a bit more insight, and then they would be able to make that prediction and reach that correct understanding. And as you contemplate the intrinsic ignorance of our species, they hit you with a 'by the way, we have discovered this incredible revolutionary thing' as they continue on their genuine Socratic plea of how little we know. As if to prove my point, Dr Kolwa continued: 'One of our big discoveries was giant radio galaxies. They've got huge radio jets that are extended over millions of light years. Very exceptional and unusual. But previous surveys were not able to see them simply because the sensitivity and resolution were not at the right level.'

These galaxies have supermassive black holes active long enough to throw jets of material for millions of years into intergalactic space. These jets emit radio waves, and from the end of one lobe to the other they are as wide as tens of Milky Ways lying next to each other. Using the word 'giant' still doesn't feel adequate enough to describe a single galaxy having such an enormous effect on the Universe.[3] And yet these leviathans were completely unseen for a very long time. Only the capabilities of the new generation of telescopes brought them into view and to our knowledge.

This is just the beginning of a deeper understanding of radio galaxies. MeerKAT was designed as a precursor of the Square Kilometre Array Observatory (SKAO), an intercontinental radio observatory with telescopes to be built in South Africa and Australia. It will be completed by 2027, heralding even more discoveries in the radio sky. I travelled to see MeerKAT and saw the beginning of SKA-Mid, the South African site of SKAO that is an extension and upgrade of MeerKAT. It was an incredible opportunity and I felt immensely lucky that I got to experience that. To get there we had to fly on a tiny plane (smaller than you think) and land on a private airstrip with an ambulance waiting at the end just in case. We couldn't have any electronic devices on during the tour, as the antennae are incredibly sensitive. That day was the longest I've been without a mobile phone in at least fifteen years. I did not have a sudden realisation that technology was evil, though. Actually, quite the opposite. I was in the midst of a nature reserve with one of the most advanced observatories on the planet, inspired by how SKAO has worked with local communities, Indigenous people and international science institutions to do the work it is doing.[4]

To Make a Black Hole from Scratch You Must First Invent the Universe

The processes that take place within and around black holes are shrouded not just in dust but also in mystery. Even their formation is not exactly clear. To make a black hole you need only have a lot of material compressed in a small enough place. I know, I'm making it sound easy, so let's add complications. First, things tend to get close together under gravity, but gravity is the weakest of the four known forces of the Universe. We experience it as a major one, but that's just because the Earth is very big compared to us 'specks upon a ball'. Electromagnetism and the strong and weak nuclear forces are much stronger. So even after gravity clumps stuff together, the other forces at the level of atoms keep things apart. Our planet can't spontaneously collapse into a black hole. To overcome those stronger forces just from a sheer amount of gravity, you need to pack a truly stupendous amount of matter in a single body very quickly. This is one of the scenarios to explain how supermassive black holes came to be: with giant clouds of gas 100,000 times heavier than the Sun, which cooled down and condensed into a single object.

The pull of gravity brought the gas together in a free fall, acquiring such momentum that it was unimpeded by other forces. In this way, the cloud turned into a supermassive black hole *seed*. Recent evidence does not favour this traditional hypothesis.[5] The preferred explanation at this time is that you still start from a big cloud but the cloud doesn't directly make a big black hole; it makes massive stars, much, much bigger than the Sun (we will talk about them in the 'Ultraviolet' chapter). In stars, gravity is counterbalanced by the energy released by nuclear fusion from the core but, eventually, the fuel for nuclear fusion runs out and gravity rules once again. The collapsing layers of the star gain momentum, eventually slamming into one another with such energy that the star explodes. This is a supernova. It would take the Sun 10 billion years to release the same amount of energy of a typical supernova. In the early Universe, these supernovae were even bigger and more powerful, creating what are now suspected to be the true seeds of supermassive black holes. In the current Universe, a supernova might end up forming (small) black holes or neutron stars, the other type of extreme object in the Universe. The difference comes from the fact that the stars used to be bigger. The elements that make us actually help in cooling down gas faster, making the protostar condense more quickly. Those elements are created in stars, so the first stars grew to bigger sizes. Bigger stars burn their fuel more quickly and end up in more dramatic explosions.

If a star that is about to go supernova is between roughly eight and twenty-five times the mass of our Sun, the energy of the collapsing stellar material is not enough to create a black hole. It will create a neutron star. However, you shouldn't be disheartened. Neutron stars are equally weird and important in radio astronomy. These stellar objects pack about 1.5 times the mass of the Sun, in a sphere smaller than most cities. The density is so high that a teaspoon of neutron star material weighs the equivalent of a mountain on Earth and the gravity is so high that mountains on neutron stars are a few millimetres tall.[6] Incredibly compact as they are, they were thought for decades to be too faint to be seen by telescopes. It wasn't until 1967 that Italian astrophysicist Franco Pacini hypothesised that if neutron stars were spinning and had strong magnetic fields, they would emit radio waves. At roughly the same time, in Cambridge, astrophysicist and all-around legend Jocelyn Bell Burnell (then a Ph.D. student) recorded – with a radio telescope she helped build – a mysterious repeated signal. That was the first recorded observation of a neutron star. It was a particular type, pulsating in radio waves, and it was given the name 'pulsar'. Pulsars are often pictured as cosmic lighthouses. The strong magnetic field creates a beam of electromagnetic radiation, and, as the neutron star spins, this beam might appear and disappear from our view, just as with a lighthouse, if lighthouses were the densest and most magnetic stellar objects in the Universe. Once again, we are walking along the shores of the unknown. Physicists from many fields are interested in neutron stars because their behaviours, properties and effects go beyond the simple astrophysical object. A neutron star is made of neutrons (the name slightly gives it away), which are particles that sit at the core of elements and have no electric charge. The properties of these particles are fairly well understood, but the jump from a neutron to an entire star composed of them is not obvious. For example, material in a neutron star is expected to aggregate itself in peculiar formations which scientists have nicknamed *nuclear pasta* because they look a little bit like pasta: spaghetti, gnocchi, lasagne, etc.[7] So, you see, astronomy truly lends itself to having recipes.

Neutron stars are a test bed for physical theories, and this role extends beyond the realm of the neutrons. One such test was about general relativity. An important prediction of Einstein's magnum opus is that a system made of two massive objects orbiting each other very closely should emit a lot of gravitational radiation. Everything with mass emits gravitational radiation but, since gravity is so weak, the effect is only noticeable when the gravitational interactions are extreme. Imagine the surface of a pond. An insect or a duck swimming would disturb the water in different ways, but both do. Thanks to this emission, the objects in this system would get even closer, losing orbital energy, until they merge. In 1974, astronomers Russell Hulse and Joseph Taylor discovered the pulsar PSR B1913+16 and its companion, a neutron star. The two dense objects orbit each other in less than eight hours. Pulsars' pulses are an extremely accurate clock – they can compete with atomic clocks – so astronomers can do amazing things with the densest timekeeper in the Universe. For example, by calculating variations in the PSR B1913+16 system, Hulse and Taylor discovered that, slowly but surely, the system is getting closer together. The system loses about seventy-six microseconds per year, getting a handful of metres closer to each other. At this rate, it would take 300 million years before they merge, and yet they are releasing an astonishing amount of gravitational energy: more than the most powerful atomic weapon ever built released in a single second. And this is just in approach! When these two stellar remnants (or two stellar-sized black holes) merge, the power would be a million billion billion times bigger.

What happens to the neutron stars is a matter for a later chapter, but let's stay on the gravitational radiation. Just like electromagnetic radiation, gravitational radiation moves as waves. We are awash with these waves, as they stretch and contract space-time all around us. But the effect is tiny, a fraction of an atom. Even to see the effects of a collision between dense objects, scientists and engineers had to build enormous laser interferometers. Within these machines, a laser beam is split in two and the two beams are sent in two different directions 90 degrees from each other. At the end of each arm, they hit a mirror a few kilometres away and they are reflected back. The setup is such that as long as the distance between the two arms is the same, the two beams should cancel each other out. But if a gravitational wave was passing, one of the arms might be slightly longer or shorter than the other. Theoretically, at least, that is the plan. But, obviously, vibrations in the mirror, in the lasers, in the building, etc. mess up things a little bit.

When I visited Virgo, the gravitational wave observatory in

Tuscany, I learned how the whole system is affected by waves crashing on the shore twenty kilometres away. The same sea of my youthful summers is now a cosmic nuisance. Researchers had to construct *superattenuators* to reduce the seismic effects (quakes, waves, cities and people) by a factor of a million billion: an engineering feat that led to the detection of a collision between small black holes, neutron stars, and even a neutron star and a black hole. As we have mentioned, everything is releasing gravitational waves, so there should be a turbulent background that permeates the whole cosmos. Our detectors are not sensitive enough to see this extremely weak signal. To observe such a signal, one would need something a lot more sensitive, a much longer detector.

How about one as big as the galaxy?

Surfing the Space-Time Waves

We have seen how pulsars are one of the subjects of study of gravitational wave research. Now we are going to discover that they can also be the instrument with which to study gravitational waves. They are not just emitters. They can also work as detectors. Radio astronomers worked out that it is possible to construct a gravitational wave detector using pulsars. This detector will not be picking out the single events like Virgo, but the constant underlying humming of our Universe: the waves of the supermassive black holes, binaries – approaching each other after their galaxies have collided – and even gravitational waves left over from the Big Bang or its immediate aftermath, a period called cosmic inflation. While the idea is extremely sound, the execution is difficult. The project is known as the Pulsar Timing Array (PTA). For the PTA to work, we need to know their pulsation timing with staggering precision. Gravitational waves in our earthly detectors create variations of just a fraction of an atom over the space of kilometres. The one being hunted with the PTA has even longer wavelengths, so you need the galactic distances. That's where the precision comes in. It allows us to monitor the space-time between us and the pulsars for gravitational waves. An extraordinary feat, and let me tell you that it is not an easy task. To find out more about it, I talked to Rami Mandow (he/him), a graduate researcher from Australia working in the Parks Pulsar Timing Array Project. Mandow explained: 'Every pulsar itself has a very incredible and very unique journey to tell us about and a very unique story to tell us about. They've got their own unique set of parameters. They spin differently at different speeds. They've got different sorts of oriented magnetic fields. They've got different binary companions.'

His enthusiasm for PTA is contagious. He is so knowledgeable and affable, sharing insight into this international work, some of which is truly mind-blowing. First and foremost, pulsars have been instrumental in expanding our idea of stellar evolution, the composition of the space between stars, and testing out our most advanced theories.

Mandow continued:

> We're kind of lucky that we have found these objects because they have revolutionised our understanding of the Universe just in the space of fifty-five years. First of all, we can learn about the intrinsic details of the neutron star itself; things like how fast it's spinning, how powerful its magnetic field is, how

> big it is, and where it is located around the galaxy. We can learn about binary companions and the systems that they're in when they're orbiting each other. We can actually use that as a laboratory for general relativity as well! Also, their signals propagate through the interstellar medium and are affected by it. So we can learn more about that interstellar medium component as well through these signals of the pulsars. So there's quite a bit we can do with them!

As I have mentioned, the signal from pulsar repeats with high regularity, but our measurements of it need to have a precision that rivals atomic clocks timekeeping, which we are close to getting. So far, astronomers have found strong evidence of a signal, but they are not yet calling it a discovery.[8] This means that the chance that detection is true and not just random noise is very high, but not quite as high as the scientists would like.

More work is needed to get to that incredible high precision demanded for the confirmed detection. That notwithstanding, the PTA is so sensitive that it can measure the mass of the planets in the solar system[9] and their effects on the barycentre of the solar system. That is the point around which all the planets orbit. It might surprise you, but it is not inside the Sun. Our star, responsible for all life on Earth, contains about 99.86 per cent of the mass of the solar system. Jupiter is slightly less than one-thousandth of the Sun but is far enough away to drag the barycentre just slightly outside our star. You don't need the PTA to work that out, but the fact that the motion of the planets can be seen in the data collected by pulsars tens of light-years away really showcases the ability of the approach and their scientists. One of the requirements for the PTA is a large number of pulsar observations. You can't do it with just a couple of them. To probe the extremely low frequencies of the gravitational wave background, you need many pulsars and you need to know them extremely well.

Mandow continued: 'By using these pulsars across the galaxy spread across the whole sky, we look for a correlated signal among them all: the gravitational wave background. And it's the only way that we can currently search for that type of signal: the nanohertz regime, the frequency of the signal itself. We need pulsars to do that. It's really important that we have pulsars!' Radio observations are crucial to it. There is a secret that is neither known nor discussed outside the field: radio waves are crucial to the discovery and characterisation of pulsars, but we do not know how those radio waves are released. It was Dr Bradley Meyers (he/him) from Curtin University in Perth, Australia, who revealed the 'secret shame' of pulsar astronomy. When I contacted Dr Meyers I had other questions in mind, but this bombshell needed to be shared far and wide, something that we joked about.

'We've known about pulsars for like fifty-plus years now. It's quite the embarrassment that we really don't know the physical mechanism that can generate the coherent radio emission that we see,' Dr Meyers explained.

Beyond the revelation, it was a fun and honest conversation; we are not under the illusion that science is flawless or all-knowing. 'There are many theories, of course, and lots of them are good at explaining specific phenomena. They can explain a handful of properties of radio emission, but they completely fail on another. Or they are really good at one thing and can't do anything else. So that's still an unsolved problem within pulsar astronomy. It's always one of these funny things that unless you're in the field, you kind of don't know. You don't realise that we don't know what is actually going on.'

I feel that this should be more widely known, not to shame pulsar astronomers, but to showcase how science is a race against our ignorance. Sometimes, even after decades, the ignorance persists. I find it very human to not know, and it is equally human to propose different solutions, even though they only partially explain what is going on. I like to see it as so quintessentially human to keep on observing and studying these objects in radio waves even if we don't know how they produce the very light that makes them discoverable to us. We will get there when we get there.

Even without knowing how this sausage is made, astronomers ought to look at their radio emission. Dr Meyers has been conducting some interesting observations. He looks at the rainbow of radio waves that you get from the pulsar. In less poetic terms: how the same objects' radio emission depends on the frequency at which they are being observed. An interesting class of objects is intermittent pulsars: neutron stars that behave like pulsars for a period, then turn themselves off, and after a while turn on again.

'Pulsars behave differently at different frequencies. Understanding how their emission, and what their phenomena look like, and comparing and contrasting them to the classical higher frequencies where most pulsar surveys happen, is really interesting. You see lots of these different phenomena that are there but maybe behave slightly differently,' Dr Meyers explained. 'I looked at a pulsar that is called an intermittent pulsar. So, it's one of those that's on and then all of a sudden it'll turn off, but it'll turn off for, like, hours or days and then come back on again as if nothing happened. I looked at that at multiple frequencies.'

Meyers found that many phenomena associated with pulsars don't care massively about frequencies. They are what we would call broadband: the general behaviour doesn't matter on frequency. That means that if it happens at one frequency, it is happening at the others, too. Well, roughly. There are some details that are important. The phenomena behave the same but do not look the same. There are specific changes that happen at specific frequencies. Dr Meyers' work has been studying at those established behaviours and seeing how they look in the radio rainbows. Who knows, maybe the answer to how pulsars make radio waves comes from a more holistic approach to their emission. As I mentioned earlier, radio waves see processes that cannot be seen in other wavelengths, making this portion of the electromagnetic spectrum crucial to our understanding of the Universe.

'Radio waves are different because they're generated by different processes. They're non-thermal processes,* and so they can reveal information about our Universe that other types of telescopes cannot,' Mandow explained. 'We can look inside and beyond dust clouds, across the galaxy and across the Universe, where optical telescopes and higher-energy telescopes can't actually pierce through. We can also study magnetic fields with radio waves, and map large-scale magnetic structures and cosmic structures!'

All these discoveries seem almost ethereal, at least for me. Take magnetic fields: they are something you 'feel', in a sense, when you play with magnets, trying to push the same poles together or pulling one off the fridge. You can visualise it with the iron-filing experiment or with a compass, but I don't immediately think about 'seeing' those fields. I'm sure some people with different

* A process that is not created by the heat released by an object.

experiences might get something more pragmatic from them, but, for me, if I need to consider just how radio waves see what we otherwise cannot, I think of the discovery of a starless galaxy.[10]

Now this sounds almost oxymoronic. A galaxy is a collection of stars, is it not? Well, sure. That's our first definition. But a galaxy has also got a supermassive black hole at its centre, and lots of gas and dust. Planets, too. And it is surrounded by dark matter. Dark matter is a hypothetical substance that doesn't interact at all with light (so technically invisible rather than dark) but interacts with gravity. So we feel it but we can't see it. It is expected that there is so much dark matter in the Universe that it outweighs the matter that makes stars, planets, us and groundhogs by five to one. What is it? We do not know. The existence of this dark matter is not in itself a great hypothesis, given how difficult it has been to prove its existence. Yet we can explain so much of the Universe and predict so much using dark matter, including the motion and behaviour of galaxies. When we study how galaxies rotate, it looks like there must be a lot more matter than we can see. It doesn't matter if the galaxies have a lot of stars or a few stars. The rotation always indicates that a lot more matter ought to be in them. Among the latter ones, there are the low-surface-brightness galaxies. These are the galaxies where most of the mass is made by dark matter (95 per cent), skewing the average a fair bit. The regular matter in them is mostly just hydrogen gas with few stars. Radio-wave observations have been crucial in studying these galaxies, since they can pick up emission from the gas between the few stars in these galaxies.

The motherlode of low-surface-brightness galaxies was discovered by accident. Researchers were doing a systematic series of observations of these kinds of galaxies using the Green Bank

Telescope, a 100-metre radio dish located in West Virginia. It's the world's largest fully steerable radio telescope and, while trying to point towards a known low-surface-brightness galaxy, researchers accidentally looked at the wrong area of the sky. A random spot with nothing in it. Or so it appears in visible light. In radio waves, they found a detection consistent with a galaxy. They called it J0613+52 and it appears to have no stars whatsoever. Just a big blob of rotating gas. The gas might be too diffuse to condense into stars, yet it's possible that stars exist in this galaxy, but a deeper search is needed to find them. The galaxy is 270 million light-years away. In cosmic terms, it is close, and yet we wouldn't have discovered it without radio waves. While unique objects are exciting, we do not believe that they are truly unique. More galaxies like this one probably exist, and radio observatories are the only ones that can find them.

Turning Venus into a Disco Ball. For Science!

Our path across the realm of radio waves has so far focused on the extreme objects, basically the very powerful, very dense, very massive. If you feel that it has all been a bit too ethereal, worry not. We are about to get more on the pragmatic side of things. We are closer to home, looking at our own cosmic backyard. Actually, let's make it even more pragmatic. We are about to deal with the length of days. As we have seen at the beginning of the chapter, we can measure the day on Earth in two ways, the solar day of twenty-four hours and the sidereal day of 23 hours and 56 minutes. The one we care about right now is the sidereal day, the rotation of the Earth on its own axis. It's easy to measure that on Earth, but how would you do it on a different world? To measure the length of a day on a different planet, you could land on it and measure how long it takes for the Sun to come overhead, but that would work for just two other planets, Mercury and Mars. For the gas giants, researchers used magnetic-field measurements and, in the case of Saturn, waves across the rings, which is a really cool way of solving this question.

Now from that list we are missing Venus. You can certainly land on Venus, but its thick cloud cover won't allow you to see much. Also, it's the hottest planet and the pressure is like being almost a kilometre underwater, so you will be crushed and melted. The thick clouds also stop measurements from orbit, which are not helped by how slowly Venus rotates around its axis. To overcome all these challenges, researchers had to get creative. They had to turn Venus into a disco ball. A disco ball is a rotating sphere covered in mirrors. Two fixed lights shine on it and the ball spins; given the angles of the mirrors on its surface, the ball moves the light across the room. If you measure how quickly the light moves, you can measure the rotation of the ball. Hopefully, you see what I'm getting at. Because researchers turned Venus into an actual disco ball using radio waves. The flashlight in question was actually a radar signal (so, focused radio waves) that could penetrate the clouds of Venus. Astronomers used the seventy-metre-wide Goldstone antenna in California's Mojave Desert to illuminate Venus – the radio waves were obviously invisible to us but the signal was equivalent to being 100,000 times brighter than an average flashlight. The signal travelled for several minutes across interplanetary space, hitting the surface of Venus and bouncing back. Part of it was reflected at a slightly different angle, so the signal was picked back up by Goldstone antenna and about twenty seconds later by the Green Bank Telescope. By combining the data, scientists were able to estimate that it rotates on its axis (its sidereal day) every 243 days, with a variation of about twenty minutes due to effects caused by its thick atmosphere.[11] A very slow but cosmic disco ball powered by radar.

Radio observations by radar in the solar system have been incredible. Even when an object doesn't emit radio waves, we can simply send a bunch of them onto its surface and be ready to retrieve them after they have travelled across the vast gulf of space between them and our instruments. Radar's obvious disadvantage is that the further you want to observe, the more power you need. It can be used to study the moons of Saturn, for example, but with current instruments you would only be able to distinguish features of about a kilometre in size. But if you look closer to home, you'll find radar has life-changing applications. Astronomers are using radar to understand the Moon as best we can before sending humans back to it and they are also using it to study asteroids which could be potentially dangerous – the so-called near-Earth objects (NEOs). To learn more, I spoke to a phenomenal expert, Dr Ed Rivera-Valentin (they/he). Dr Rivera-Valentin is a planetary scientist at Johns Hopkins University Applied Physics Lab, and our conversation spanned our shared interests, from Pokémon and bow ties to the planet-saving abilities of radio observatories, including the defunct Arecibo Telescope, which was located in Puerto Rico (where Rivera-Valentin is from) and appeared in films such as *Contact* and *Goldeneye*.

'Radar is applied to facilitate planetary exploration. So all of it comes together through this applied-science lens of being able to facilitate safe exploration of the solar system and to safeguard ourselves and protect the Earth from devastation … uniting all peoples within our nation,' they told me, channelling Pokémon's Team Rocket. Dr Rivera-Valentin is referring to the risk of dangerous asteroids colliding with our planet. Fortunately this does not include Asteroid (389478) Rivera-Valentín,* which is named after them. Dr Rivera-Valentin's expertise spans several fields. They work on the habitability of other planetary bodies like Mars, as well as studying the potential contamination risks from our robotic explorer. Their other area of expertise is radar observations, especially of asteroids and the Moon, including work that will help the work astronauts will do when they go back on the Moon with the NASA-led Artemis mission this decade.

'One of the projects I've been working on is studying the current Artemis candidate landing sites. Actually, across several of them, you see this very long ray coming from a distant crater in the radar, but you don't see it in other wavelengths,' Dr Rivera-Valentin explained. 'If you have an astronaut going there and they're going to pick up a rock, we want to make sure that we know what's the probability that that rock is actually from that area versus they're sampling something that came from the equator or something from a large crater that ejected all this material over.'

The highest-precision images of the Moon from the surface of Earth were done using radar. With observations that can reach a resolution of five metres per pixel, we are reaching the precision of the spacecraft actually orbiting the Moon. And this was in a test for this radar technology. The power used in this

* You can't just ask to have an asteroid named after you, unfortunately. I would have asked by now, otherwise!

test is roughly that of a common household appliance, just 700 watts. With a future plan, to make it almost 1,000 times more powerful, researchers expect an incredible level of insight in our study of the surface of the Moon. Well, actually the subsurface. You see, the radio waves can penetrate the regolith, the lunar soil, down to a few metres.

'If you want to get that historical planetary geology perspective, you want to study the subsurface. So unless you're going to send people there to be digging or you have a crater that exposes that stratification, one of the best ways that we have is to do things with radar,' Dr Rivera-Valentin argued.

A great example they put forward is Mare Crisium, or the Sea of Crises. This is not a nickname for the political landscape of the last decade, but one of the dark basaltic plains that cover the near side of the Moon. You can see it with the naked eye: it is the roughly circular one on the top right (north-east) of the Moon's face. Planetary scientists have used different radio wavelengths (known as S-band and P-band) to observe this famous lunar sea with radar and discovered a young crater and old lava flow, something that would not have been possible with visible light alone from orbit.

Dr Rivera-Valentin further explained:

> There's these really two great images of Mare Crisium on the Moon in radar. One of them is in S-band, so that's twelve-centimetre wavelength. And the other one is in P-band, which is seventy-centimetre wavelength. So the S-band sees about down to a metre while the P-band sees seven to ten metres. In the S-band, Mare Crisium looks kind of like a little Christmas tree. All the fresh craters show up really bright. But

in P-band, you're not really seeing those fresh craters; instead, you end up seeing these ancient lava flows coming out from one of the edges of Mare Crisium.

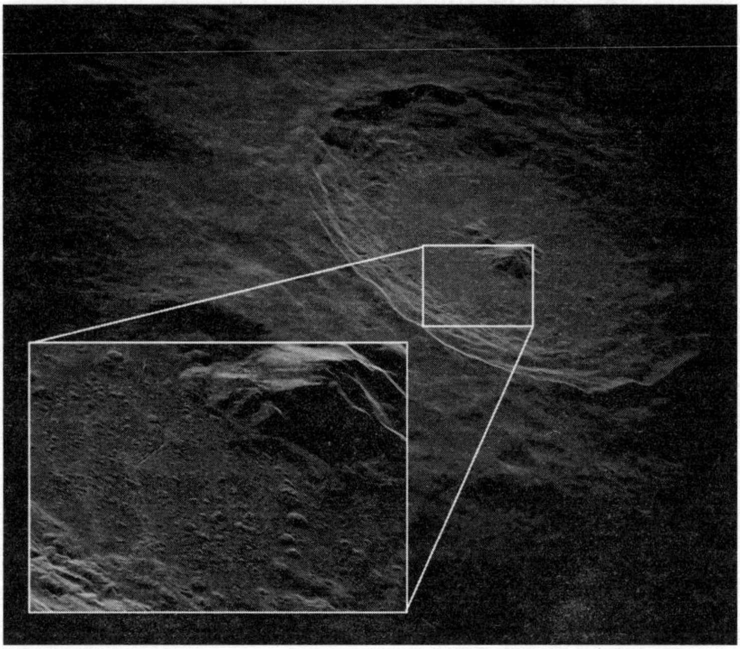

The radar image of Tycho's crater, showing a five-metre resolution detail

Raytheon Technologies

The radar observations of the Moon allow scientists to look back in time, discovering new insights into known locations with the potential to alter the future plans for permanent bases on the Moon.

Armageddon and Deep Impact Meet Reality

Planetary defence arouses complicated feelings in me. On one hand, we have confidently found all the near-Earth objects that could wipe us out completely. That is good news. Thanks to NASA's DART, we have demonstrated that we can deflect a dangerous object, given enough time. Also excellent news! So why do I feel concerned? Because we are still missing many of the smaller objects, those that could bring devastation on a local scale. Far too often we spot NEOs just days before they fly by.

And at the time of writing, only eleven asteroids (all so small that they burned up in the atmosphere) were spotted ahead of their impact on our planet.[12] Eleven: a tiny number! The crucial problem is that asteroids are small and cold and the Universe is big and, in general, also pretty cold, so you tend to spot them when they get close enough to Earth. If you spot them only once, you have massive uncertainties, so you need multiple observations and you might still not be very sure because visible light is never quite precise, even up close. These NEOs are really dim. And uncertainty kills. That's when radar's utility kicks in. You are sending radio waves to these objects and you can do two quick and important types of science.

Dr Rivera-Valentin gave me the lowdown on the true power of radar:

> The first one is that you can immediately nail down the orbit of the object. With, for example, Arecibo [Observatory] in S-band we were able to get the distance of the asteroid down to about a metre and the velocity down to dozens of centimetres

> per second. That precision enables you to quickly put together the orbit of the asteroid and be able to predict where it's going to be out to 100, 200 years in the future. To get that level of precision, you would need many nights of optical observation. But with radar, give me like a couple of minutes on this thing and I got your orbit down.

As mentioned earlier, the Arecibo telescope is no longer operating. It broke apart a few years ago, about which Dr Rivera-Valentin and I commiserated with each other during our chat. There are other instruments, but planetary defence needs investment if lives are to be saved. If impact is imminent, having radar observations would allow scientists to work out where exactly it will be and prepare for evacuation. There are actually exercises by space agencies and other organisations to prepare for such eventualities. If an asteroid is on a dangerous orbit with a possibility of impact with our planet in the near future, then the precise measurements of its orbit will be crucial in order to plan a deflection mission. But knowing the orbit alone is not sufficient. It's important to know what the asteroid is like. And, once again, radar is our friend here.

Dr Rivera-Valentin further explained:

> The other aspect that radar gets you is this interesting, nice image of the asteroid. I can put those images together to tell if this is a spinning-top-shaped object like Bennu, or if it is a dog-bone-shaped object like Cleopatra. From the surface, I can figure out if it is a rubble pile: what's the density of the surface? How many boulders are all over the place? Is there a crater somewhere? I can do that physical characterisation

> of the surface that's needed to then plan on that mitigation. Because if I'm impacting a rubble-pile asteroid that looks like a diamond versus a monolithic, very strong object that looks like a dog bone, the type of mission I would put together is going to be different to push it away just right so this object ends up missing us by the time it might come towards Earth.

Radar is not just a crystal ball into the future when it comes to asteroids, it is also a magic mirror on the wall that tells you what they are like. That said, I don't think radar has been tested to look at the fairest of them all. Asteroid Bennu is pretty, but beauty is in the eye of the beholder. It is certainly the most dangerous currently known asteroid. There is about a one in 1,750 chance of it hitting Earth by the year 2300.[13] The most likely date of impact, where the orbits are predicted to be closest, is 24 September 2182. Mark your calendars.

Desperately Seeking Someone … Anyone?

A chapter on radio waves would not be complete if we did not consider a trifling question. Are we alone in the Universe? To paraphrase the legendary Miss Britney Spears, our cosmic loneliness might not be killing us, but it is a constant line of enquiry. From science luminaries to UFO cranks (and the surprising overlapping of the two – feel free to email me for the tea on that), people's hopes, beliefs and hypotheses end up facing a Universe that is not willing to give us a straight answer. At least not just yet. I'm going to be upfront and say what I believe and what has informed such belief. I do not claim that I am right – and if you have spent time around physicists, you know that is a big thing to admit. I will, however, claim that it is reasonable, and that it is not merely a flight of fancy.

I believe that life is common in the Universe. Not everywhere, but common nevertheless. There are multiple worlds in the solar system that might have had (at one time or another) the right conditions for life. We have several icy moons around the gas giants with deep liquid oceans and hydrothermal structures that can provide the chemistry and energy for life.[14] Planets like Mars had a watery and volcanic past.[15] Maybe even Venus had an ocean.[16] The right conditions do not make life a certainty, but if we are correct in our understanding of how life developed on our planet, it is a good thing that we see the same ingredients and processes at work in so many different places. So my main belief is that if we are to find life, it is going to be in the shape of simple life forms.

A discovery of this magnitude would be epochal. We could understand so much more about life on Earth. Is DNA the only way to make life? Are there strict conditions in order for life to evolve, or is there flexibility? We would know about so much more than we know now. But even the more enthusiastic biologist and astrobiologist might be left a tiny bit unsatisfied. I think our cosmic loneliness and our search for other intelligent species out there stems from our search for purpose. Why are we here? Why on this beautiful planet full of different (and some quite intelligent) species do we alone wonder about the mysteries of the cosmos? An answer might come from encountering another extraterrestrial civilisation: it would tell us that consciousness and curiosity might just be common traits, that we are all instinctively looking at the sky, reaching, or wanting to reach, the stars.

I believe there are others like us out there, but maybe too few of us to get talking straightaway. After all, we have only been sending signals out into the galaxy for about a century, and if that isn't enough, every time the distance doubles, the strength of the signal is reduced by a quarter. Our signals have only reached stars within 100 light-years, assuming they are still detectable at those distances. The entire galaxy is more than 1,000 times wider than that. The other side of the argument is that there are around 60,000 stars within that distance from Earth. In the words of Italian physicist Enrico Fermi: *where is everybody?*

Fermi's question came in 1950 during a lunchtime discussion with colleagues on UFOs and faster-than-light travel. It became known as the Fermi Paradox. We do not know if faster-than-light-travel is possible, and interstellar travel is doable but only with ample resources. A civilisation could easily spread among stars to settle on new planets, and spread throughout the galaxy over many hundreds of thousands of years. We are not there yet in terms of technology, but we could easily do it without having to develop things beyond our current understanding of physics. It might be possible.[17] But apparently nobody is doing it. Or at least we have no clear evidence that any civilisation is doing it. Maybe interstellar travel is too costly or simply not worth it. The same goes for creating incredible structures.[18] Or perhaps intelligent life is just too rare. After Fermi set his paradox, it took another decade for scientists to start looking for a concrete answer to his question. That quest for a definitive solution to the Fermi Paradox is what we call SETI, the search for extraterrestrial intelligence. To get insights from the past, present and future of SETI, I went directly to the SETI Institute.

Dr Franck Marchis (he/him), senior planetary astronomer at the Institute, was ready to shower me with insights about the hunt for alien civilisations.

> SETI is a field of research in astronomy and the goal of SETI is to search for technosignatures or proof of intelligence in the Universe. And there are multiple ways to do it. The most famous method [employed by SETI], the one that most people know about, is using radio waves. That started in the sixties when Frank Drake started Project Ozma, where he pointed an antenna to two nearby stars. His idea was that if there is a civilisation on planets around those stars and those civilisations are identical to ours, then they will probably evolve like us. So they will have radios, radar, and so on. So we should basically point super-powerful antennas towards those planetary systems. And if there is a planet habitable and a civilisation, we should hear them. That's the idea. That's the radio way.

Radio is not the only way to communicate across the distance in space. Humans, too, have started to use lasers from spacecraft to send data, images and, of course, a cat video![19] So we might spot those lasers coming from other places in the galaxy. Or maybe the communications are even more sophisticated, using quantum mechanics. If we can transmit information, maybe aliens are using it as their primary mode of communication; we could spot them like that. A less futuristic approach is detecting technosignatures in the atmosphere of planets, from chemical signals or even the heat of cities, something we will touch upon in the chapter on infrared.

While other avenues are being investigated as they should be, radio waves are not being abandoned. In a way, we can consider that looking for alien civilisations in radio waves is just getting started. The search for SETI messages over the last six decades has not been an all-encompassing affair. Prior to that, technological limitation meant that only transmission directed towards us would be picked up. Now, airport radars, mobile-phone masts and every radio transmission that escapes into space make our planet radio loud. Other worlds with intelligent civilisations might be like ours, with TV, aeroplanes and mobile phones. And soon we might be able to spot them. I've already mentioned that the next big thing in radio observatories is SKAO. An alien civilisation with a SKAO could detect our mobile-phone network from ten light-years away.[20] But it's not just new tech, it is also new ways of using the old tech. Researchers have now developed a way to scan through the data that the various telescopes collect more effectively,[21] even if they are focusing on other things, allowing them to go through 250 times more sources than a star-by-star approach.

'I think it's important we continue to search in radio because we have been nosier in radio for much longer than we have been in lasers. Plus, it doesn't cost much. I mean, a radio telescope costs almost nothing compared to a space telescope,' Dr Marchis continued. And here, too, we see the versatility of radio telescopes. Because SETI might not have found aliens yet doesn't mean that those searches are useless. One excellent aspect of science is that even finding nothing tells you a lot. These searches have not found a radio-capable extraterrestrial civilisation within a certain distance from Earth, but they have seen a lot of other things, including some truly weird stuff. As we are about to see.

A Flash in the Disc

About a decade ago, scientists looking at pulsars made a fascinating discovery. Among a survey on pulsar emissions, they saw a brief pulse of energy in radio waves. This, known as the Lorimer Burst, would go on to create a whole new field of interest in radio astronomy. The object of study of this field is none other than fast radio bursts (FRBs), short and powerful emissions of (mostly) unknown origin. I won't pretend that I do not have favourite discoveries in astronomy. First, because it is normal to have preferences, and, second, because I would be lying if I said I care about everything else in radio as much as I care about FRBs. They are both awesome and a big puzzle. What's not to love? But let me make a good case for them and, hopefully, you'll love them, too. Or at least see where I'm coming from.

The best clue for the source of these events is in how much energy they release. These FRBs are extremely powerful but incredibly brief flashes of radio waves coming from distant galaxies and lasting just milliseconds. And in that fraction of a second, they release energy equivalent to what our Sun might produce in days. Now that is a lot of energy to be released in such a short time, but it is also so much smaller than what an event such as a supernova might release in minutes. We have seen that a supernova releases the equivalent of 10 billion years of solar energy during the explosion, so we are looking for something powerful but not as incredibly powerful as a supernova.

Asking Dr Meyers about them, his answer backs up my feeling that they are awesome (and remember, he's a pulsar astronomer):

> They're sort of mysterious, they're more mysterious than pulsars. And we still have a lot to understand about pulsars. I mentioned we don't even know how pulsars emit radio waves. The same is true for FRBs, but they're sort of the next level because they're in other galaxies and we don't really know their origin. And there's all this wonderful physics that you can do with FRBs that you can't do with pulsars. One of those things is you can do cosmology.

The last statement seems almost a throwaway line. What is cosmology? The study of the Universe as a whole. Seems vague enough. So let me put it in more pragmatic terms and give you an example. Until 2020, we didn't know where half of all the regular matter of the Universe was located. This matter is what is known as baryons and makes you, and me, this book or ebook reader, planets, platypuses, stars and comets. Using our best understanding of the Universe, the Standard Model of cosmology, we worked out how much baryon mass exists in the visible Universe. Unfortunately, we can see only half of it. That half is in stars and galaxies. The missing baryons have been assumed to be spread out, in extremely hot plasma across intergalactic space. This is too faint and too diffused for our instruments to detect. That's where the FRBs come in. Their emission is a perfect probe for the matter that exists between galaxies. The radio waves from FRBs get spread out as they pass through the intergalactic medium, just like what happens to sunlight in a prism. But this invisible radio rainbow confirms that there is a lot of matter in a hot and diffuse plasma between the galaxies. We just had to discover a whole new and unrelated phenomenon to find a solution to this mystery.

If all this has not sold you on FRBs yet, let's continue with what we believe to be their origin. Again another connection to pulsars. Some FRBs repeat, others do not. Some seem to be from a clear type of source, although many different events could explain others. Magnetars – a version of neutron stars with incredible magnetic fields – are considered a possible progenitor of some of these events.

'It seems like a compact object origin is the only thing that can produce energies enough to create these enormous bursts that are seen from across the Universe. That kind of thinking led us to consider "oh, well, maybe it's a neutron star", and pulsars are neutron stars,' Dr Meyers added.

The idea is that these magnetars are orbiting another object and, regularly, the conditions are right for a release of radio waves. What the exact conditions are is a work in progress. Still, astronomers are using the know-how from pulsar observations to better understand FRBs. But that's not all. We ended the previous section by talking about how SETI searches deliver so much even without finding aliens, and a fantastic example is FRBs. Recent observations by SETI did indeed deliver more insights into fast radio bursts,[22] even though we did not hear from any little green men. We can actually create scaled-down versions of these FRBs at home. You just need to open the door of a microwave before the timer goes off and, bam, there it is: a miniature fast radio burst. Signals like that are now called perytons and they did appear in actual astronomical observations, but they were recognised pretty quickly as coming from Earth.[23] Given how much energy an FRB releases, our microwave analogy gets us only so far.

But electronics can accidentally make it appear that they are alien signals. Back in 2020 a radio signal from the closest star to the Sun got people excited. Proxima Centauri hosts planets, so could it have a technological civilisation? At first glance, the signal was promising. It disappeared when looking away from the star, which is what you want. It seemed to be coming from a specific place in the sky. It looked like the most promising alien signal since the Wow! signal – a strong radio signal that was detected in 1977 and then never again. Extraordinary claims demand extraordinary evidence, so the team searched repeatedly for a possible explanation. And they found it. The same signal was found in other directions and by other telescopes. The radio frequency is emitted by oscillators found in several common electronic devices, although they couldn't work out which specific ones. The search is difficult but it is rewarding since, as I've mentioned, we can learn about so many other aspects of the radio Universe, even if we have yet to find aliens.

For Marchis and many others, the ability to discover more than expected shows the enormous impact of SETI. The driving force remains not just finding aliens but talking to them. 'I mentioned multiple times in my talks that I don't care whether or not we are alone because I don't believe we are alone. What I care about is to have meaningful communication with aliens. That's way more important! Because having meaningful conversations with aliens will give us a lot of information about ourselves, help us overcome problems, and so on … it is going to be way more interesting!'

The Only Light in the Dark

This chapter should have showcased just how incredible radio astronomy is but, to be fair, I have barely scratched the surface of the way we use it and the cool stuff we get from it. But there is an even more impressive feature of the Universe that only radio astronomy is able to deliver. And, yes, it is cooler than starless galaxies, dangerous asteroids, and even aliens. That feature is the second light to shine through the Universe. Radio waves can literally let us go back in time almost as far as we can see.

To discuss that special light, we need to establish a quick chronology of the cosmos. The Universe began its existence with an event called the Big Bang. For about 400,000 years, it was so hot that atoms were unable to form. As the cosmos expanded, it cooled down and all the hydrogen and helium and a sprinkling of lithium was finally able to hold on to their electrons. But something else happened, too: the heat from the Universe was another form of light and suddenly that light was free to move. (That light is known as the cosmic microwave background – CMB – and we will learn all about it and the aftermath of the Big Bang in the next chapter, on microwaves.)

Going back to the hydrogen that had just formed, the whole Universe was a fog of this gas. We do not naturally think of 'fog' as a source of light, but there is a particular emission from hydrogen that we can measure using radio waves. This is the second light in the Universe. A hydrogen atom is made of one proton and one electron. Occasionally, we get a proton and a neutron or two neutrons in the core of the hydrogen. These versions are isotopes of hydrogen and are called deuterium and tritium, but they are just a tiny fraction of the total hydrogen in nature.

Particles possess an *intrinsic* form of angular momentum that we refer to as spin. We can picture each particle like a little spinning top. It is not exactly correct, but we are not doing a test at the end of this, so the analogy is good enough for us. In a hydrogen atom, the spin of the proton and electron can point in the same direction (both 'up') or they can point in opposite direction (one 'up' and one 'down'). This means that if they were spinning tops they would both be spinning clockwise in the first example. The lowest possible energy state is when they are pointing in opposite directions. So if they are not, the system has a little bit of extra energy. The Universe hates things that have a bit of extra energy; it likes stuff at the lowest possible energy – the ground state – so, in time, that hydrogen atom with the same-direction spins will release some energy and perform a so-called spin-flip. The energy is lost by the atom, following the emission, as a photon with a specific energy. An energy corresponds to a specific frequency and wavelength, which leads to the name of this emission line: it is known as the 21-centimetre line.

Twenty-one centimetres is actually in the microwave range, but the emission is studied in radio waves. This is due to a crucial feature of the Universe. The cosmos is experiencing an accelerated expansion. At large scales, everything is moving away from everything else. 'Large scales' are the key words there. Your home is still the same size, day in and day out. During the pandemic lockdowns, you might even have felt as if it was getting smaller. We do not see the effects of the expansion on a human scale or even a galactic scale: you need to go to extragalactic scales. That's how we know about expansion – the light we get from galaxies changes subtly.

Imagine yourself by the side of the road and an ambulance is approaching you, siren blaring. As it comes closer, the sound will increase in pitch; once it passes, the pitch becomes lower. This is the Doppler effect and it's due to the motion of the ambulance compared to you, the observer. The sound waves have been compressed in approach and stretched out as the ambulance swooshes away. A source of light moving towards you or away from you will also have this effect. Just as with the sound, the light of the ambulance has shifted but it is not moving fast enough for humans to appreciate this effect. You need to go at tens of kilometres per second to start noticing, so if you are moving at those speeds, the light from an object approaching you will be shifted to the blue end of the visible spectrum, and the light from an object going away from you will instead be stretched into the red end. So the light of every galaxy (bar a few local ones) is redshifted, meaning that they are moving away from us. But it is not a real Doppler shift. The galaxies might be moving, but most of the redshift we measure is due to the expansion of the Universe that makes it look that way. The emission from the 21-centimetre line is the same. It will be in the microwave range, but far away from Earth it gets stretched into longer and longer wavelengths that we observe in radio waves.

There are two reasons why we are interested in this emission. One reason is to find holes in the fog of hydrogen. The holes mark the moment when the hydrogen is once again stripped of its electron, a period we call reionisation. The cause of the Universe becoming transparent is the formation of the very first star inside the very first galaxies. Their light is energetic enough to rip the electrons off hydrogen atoms. To better understand it, I had to find someone who hunts this light for a living. That person was Aman Chokshi (he/him) who was at the University of Melbourne when we spoke. Chokshi's area of interest is the epoch of reionisation, and his radio astronomy took him all the way down to Antarctica to use the South Pole Telescope, something I am very jealous of, as I make clear in our chat.

'The epoch of reionisation is the origin of this important light in the Universe,' Chokshi explained.

> After the Big Bang, galaxies formed in dark matter halos. The birth of the first stars, galaxies, black holes led to the emission of ionising radiation, which propagated in spheres … spheres relatively close to these halos for a time. As these spheres of ionising radiation propagated, they ionised all the hydrogen they pass through. The Universe transitioned from neutral to ionised in these bubbles. Eventually, these bubbles start to overlap, and over time [all the hydrogen] is ionised.

The whole affair probably took 850 million years. The first stars and galaxies might have begun to form around 150 million years after the Big Bang, and by the time their light ionised all the hydrogen in the Universe, our cosmos was celebrating its one billionth birthday. Using radio waves, astronomers hope to find the signals from the bubbles of ionised hydrogen at different epochs, providing insights into the Universe in its infancy. It's thanks to these kind of observations that we are beginning to understand a period that has been aptly called the Cosmic Dark Ages.

The other reason we're interested in these emissions is that they enable us to understand the Universe on a much wider scale. By mapping the emission across many redshifts, we can reconstruct the distribution of matter over time. This is a goldmine for answering questions about the nature of dark matter, for example, and other unresolved questions in cosmology.

Chokshi continued,

> By identifying neutral hydrogen at different wavelengths, we can tell exactly which epoch of the early Universe the light was emitted at. Essentially the idea is that if you can detect this 21-centimetre light at a range of frequencies across cosmic time, you have the possibility to tomographically map out the history of and the evolution of neutral hydrogen over this cosmic time. So you can form a 3D image, slice by slice, at different frequencies, mapping out the neutral hydrogen as it turns into ionised hydrogen as we get closer to us.

The seed of that distribution is already visible in the cosmic microwave background which is now upon us as we move to the next chapter. But that is a flat image, a snapshot. The 21-cm line observations give a three-dimensional map of the universe, detailing how things changed during one of the most mysterious periods of the cosmos. This is terribly exciting but also terribly difficult. The signal from the neutral hydrogen is weak, much weaker than human interference and the emission of a lot of other celestial bodies. But SKAO, the incumbent new kid on the block of radio astronomy, is expected to provide the level of precision necessary to bring that signal to light.

Our first foray into the invisible Universe has shown you that there is light unrelated to the temperature of an object or how much it shines. You have become acquainted with a large cast of characters, which we will see again and again on this journey.

But we know now that they might not look or behave exactly the same way in other wavelengths. Our transformation has begun. Radio astronomy shows us the stunning breadth of possibilities of what lies beyond our visible spectrum. We have left what we can sense behind and we are meeting a Universe of extreme objects. A Universe of powerful events. A Universe of changing galaxies. A Universe of dangerous asteroids. A Universe that appears silent but, maybe soon, we will hear that it is teeming with life.

Radio waves are the key to those discoveries and, historically, the successes of radio astronomy spurred the use of other wavelengths. We are on the same road that countless astronomers and the field as a whole took before us. Radio has opened a wide window into the previously unknown. There is so much more still to explore. But before we do this, let us take one last big gulp of air here ... we might just get a sense of a cloud that tastes of raspberry and smells of rum.

— CHAPTER 2 —

MICROWAVES

The Bigger Picture

> In the beginning the Universe was created. This has made a lot of people very angry and been widely regarded as a bad move.
>
> —*Douglas Adams*

> I saw no Way - The Heavens were stitched -
> I felt the Columns close -
> The Earth reversed her Hemispheres -
> I touched the Universe -
>
> —*Emily Dickinson*

Throughout this book we have been treating the wavelengths of light as fitting into neat little boxes, but the spectrum is continuous. Microwaves are not that far removed from what we do in radio waves and infrared. But even so, microwaves deserve their own chapter for one specific reason: the fundamental piece of evidence that the Universe began with the Big Bang comes from the cosmic microwave background. Our journey into these wavelengths will discuss the eponymous 'bigger picture' in detail, but there are other wider understandings of astrophysical phenomena that come from this range of light. Think of the frescos of the Sistine Chapel, or Hieronymus Bosch's painting *The Garden of Earthly Delights*. Think of the book that you like to read and reread. It's about the broad brushstrokes as well as the details. And this is the Universe in microwaves. The bigger picture. And the details that will change your understanding of it.

It is wonderfully apt that I am writing about the very beginning of the Universe in the beautiful theological room of Gladstone's Library in North Wales. Organised religion and I have a rocky past, ever since I was kicked out of Sunday School for pointing out that the Big Bang Theory is the accepted beginning of the Universe by the Catholic Church. (I was right, but adults do not like to be corrected by kids.) The question of *what* sparked the Big Bang is a matter of belief, but from a fraction of a second afterwards we can follow its evolution with physics. Let's rewind the clock by 13.8 billion years. Back then, everything was in the same place. The Big Bang happened everywhere. There is no centre of the Universe. It all started at every point. This is already challenging our little brains, but we will need to go further, because, at the instant of the Big Bang, both space and time fail to make sense. All our known physics fails. And it is fascinating to understand why. To appreciate that, we need to know a bit about fundamental physics, so here is a crash course.

In the last chapter we saw that the matter of the Universe can be divided into two: dark matter and baryonic matter. Dark matter is a mysterious and invisible form of matter and its existence has yet to be confirmed. Baryonic matter is what we are made of. It's what atoms are made of, plus a load of little weird stuff that is unfortunately very important and can't be ignored. You are made of different types of atoms. The type of atoms depends on the number of protons in the nucleus. Protons are particles with a positive electric charge. For example, hydrogen is an atom with one proton. Two protons make helium. Three is lithium, and so on. The nucleus of atoms also has another particle called a neutron that is there to make sure protons do not push away from each other. Particles with the same charge repel each other, like trying to make the same poles of two magnets touch. The last component of the atoms is electrons. Usually there are the same number of protons and electrons in an atom.

It turns out that electrons are fundamental particles. This means that scientists have smashed them together in a lab and found out there's nothing inside them. That's the end of the line, they're fundamental, there's nowhere else to go with them. I've heard many jokes about particle physicists being like children smashing things together to see how they work and giving them funny names. We are about to meet particles with 'flavour' (and that has nothing to do with taste) as well as the 'gluons', so-named because they make things stick together, so there might be a grain of truth in those jokes. Still, being childish is fun, especially in serious contexts, and I'm all for it.

Protons and neutrons are not fundamental particles. When physicists smashed them together, they found something else inside: quarks. Quark is a portmanteau of question mark (from James Joyce's *Finnegans Wake*), an extremely appropriate name given their uncertain nature when they were discovered. There are six types of quarks. Protons and neutrons have 'up' quarks and 'down' quarks, and their stability is the reason why we are here; they don't decay quickly. The other four types of quark – charm, strange, top and bottom – are rarely seen (only after very energetic events) and as a result they are not stable. These six species are referred to as the *flavour** of quarks. You do not need to lick a top quark to know it's a top.

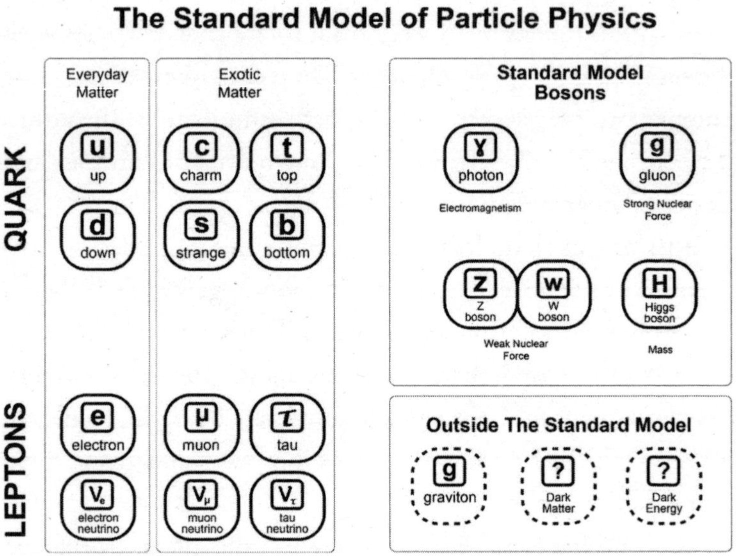

This infographic shows in simplistic terms the Standard Model of particle physics and what's beyond it.

Alfredo & Chris Carpineti

Now, I mentioned that neutrons are in the atoms to stop protons from flying off due to their positive charge. These neutral particles are able to do that thanks to gluons, the sticky mediator of the strong nuclear force. This is one of the four fundamental forces of the Universe, and it only operates inside the nucleus. So it doesn't really matter to us, apart from the fact that its

* The term 'flavour' was invented in 1971 by Murray Gell-Mann and Harald Fritzsch, while sitting at a Baskin-Robbins ice-cream store in Pasadena, California. Quarks, just like ice cream, have both 'colour' and 'flavour'.

existence and its strength made us possible. We only experience two of the four fundamental forces: gravity (mediated by the yet-to-be-observed graviton) and electromagnetism, which is mediated by the photon, connecting light to the fundamental forces. To discuss the last fundamental force, we need to know about the flavour of the electron. Just as there are six flavours of quarks, there are six flavours of leptons, a class of fundamental particles that include the electron. The muon and the tau particles are heavier cousins of the electrons, and then there are the three neutrinos, called the electron neutrino, the muon neutrino and the tau neutrino. The remaining fundamental force, known as the weak nuclear force, can change the flavour of some of these particles. It's responsible for some nuclear interactions and causes radioactivity in ways that are well beyond the scope of this book. It is also behind the fusion processes that power the stars, so let's stress that it is important.

Now, all these different flavours of fundamental particles have a mirror version called 'antimatter'. For the electron, the particle is called a positron. This is not just an electron with a goatee and evil intent. It has the same mass but the opposite electric charge. And if matter and antimatter meet, they go *boom* and turn into pure energy. This is called annihilation.

I want to stress the importance of neutrinos. These tiny particles won't feature much in the book, but they are produced in many astrophysical events. Every second, 61 billion neutrinos produced by the Sun fly through every square centimetre of your body and you are none the wiser for it.[1] They are electrically neutral, as the name suggests, and they have a mass so tiny that for a long time we weren't really sure they were there at all. And they have a property which is absolutely bonkers: while the flavour of the other quarks and leptons are stable unless they're changed by an external force, neutrinos oscillate between flavours. As they travel through space they can switch from one flavour to another, summarised in the words of Dr Linda Cremonesi that neutrinos are *flavour-queer*. Love this journey for them. Together, all these particles and forces make up the Standard Model of particle physics: one of the best physical theories ever conceived! I am not exaggerating. If we needed to show off in front of aliens I'd be bringing the Standard Model, the frescos of the Sistine Chapel and 'Bohemian Rhapsody'.

The Standard Model is such a magnificent description of the fundamental Universe, and I need you to appreciate that, because we're about to shit all over it.

The Limits of Our Physics

The Standard Model is great but it is limited – it struggles to deal with gravity, it doesn't include dark matter and it doesn't include dark energy. Radio waves show us that the Universe is expanding and that this expansion is actually accelerating. To explain this, physicists have proposed something else that is invisible (again with the dark misnomer) – an energy that affects the Universe at large scale. Dark matter and dark energy make up 95 per cent of the energy content of the Universe, regular matter the remaining 5 per cent. So the Standard Model only explains about one-twentieth of the cosmos …

I wish that was all. Gravity doesn't fit too well into the model. Our theory of gravity is Einstein's general relativity and it has passed every test we have thrown at it. It is great. The other forces are explained by the Standard Model and quantum mechanics, and they have also passed every test (although we might be seeing some tiny issues). You'd think that combining the two most incredible physical theories would give you an even better one. What it gives you instead is nonsense. You put the two together and they don't work. There are two places where we need them to work and they don't: inside black holes and right after the Big Bang. As we approach the instant we like to call the Big Bang, our laws of physics change. The fundamental forces begin to unify. The temperature is so high, the energy of reality so outstanding, that the division between forces is meaningless; we are fractions of seconds before the beginning, and electromagnetism and the weak force become one. As we get closer, the strong nuclear force joins in, and truly near the very instant that our cosmos started, gravity is there as well. There are no particles there and at that point our physics is meaningless.

The word 'incomprehensible' is perfect for this. First, because we cannot use the science we hold dear to understand it. Second, because it doesn't work in our heads to think of a moment when all science breaks down. Physics begins to make sense about one billionth of a second after the Big Bang. At that point, the fundamental forces are all separated. Fundamental particles are also beginning to pop out around this time, but they are still in an extremely hot plasma. The work done in particle accelerators such as the Large Hadron Collider is to produce these conditions that existed for a fraction of a second. By the time we reach one second after the Big Bang, we believe two things have happened. One, neutrinos are not stuck with the rest of the matter, so they begin to travel freely (something known as the cosmic neutrino background). Two, matter dominates. That's the crucial fact to explaining why we are here today. During the early phases, matter and antimatter were constantly popping in and out of existence. Energy can spontaneously turn into a particle-and-its-antiparticle pair and that pair interacts, going back into being energy. Something in the Universe favoured matter over antimatter, and a little bit of the matter survived the process. We are the descendants of the matter the Universe couldn't destroy and the Standard Model can't tell us why.

From that first second to the first twenty minutes of the Universe's existence, the Universe continued to expand and cool. The two go together, just as a spray can gets cooler as you use it (there's less gas inside, so more volume for the gas to expand into). Across those minutes, the plasma cooled down and the quarks combined in the protons and neutrons. The nuclei of the first atoms began to form. Of all atomic nuclei, 75 per cent were hydrogen - a single proton - and almost all the rest were helium - two protons and two neutrons. There was also a sprinkling of lithium. As we saw at the end of the last chapter, the atoms without electrons were ionised. The Universe was just too hot. Photons were bouncing about with so much energy that every time an electron got together with a proton, they would kick it out. The Universe had to get much cooler and so much larger for matter to form. For about 400,000 years photons had too much energy. It did not all happen at the same time – there were small variations – but eventually the original photons from the Big Bang were free to travel. Those photons are still with us today, and they are the main subject of this chapter. The Universe has expanded and cooled a lot more since then.

The cosmos is now 13.7 billion years old. Back 400,000 years after the Big Bang, that light would have been visible to our eyes ... if eyes had existed. It is now in the microwave realm. It is the cosmic microwave background, the echo of the Big Bang. It is also the furthest we can ever hope to use light to study the Universe. There is no free light beyond it, a hard limit to our past. It is the first light. It is almost insulting (but very human) that its discovery had to come through pigeon crap. Talk about humble beginnings.

The Echo of the Beginning

I feel I should probably apologise for the several mentions of religious dogma and faeces we have in this chapter. Make of that what you will, but we are not out of either quagmire just yet. As modern physics was being constructed at the beginning of the last century, scientists rejected the idea of 'creation' by embracing another type of dogma: the steady-state Universe.

Astronomers believed the Universe always existed and remained unchanged. But a schism was coming. In 1929, Edwin Hubble, using distance methods developed by astronomer Henrietta Swan Leavitt, showed that distant galaxies were moving away from Earth. The more distant they were, the faster they were moving away from us. The Universe was expanding, getting bigger in every direction. This fact meant that if you were to run it backwards it would all end up collapsing into a single event that exploded outwards. For this reason, steady-state supporter and Cambridge astronomer Fred Hoyle *mockingly* called this theory 'the Big

Bang'. Hubble was right, though, and Hoyle wasn't. What was missing was direct evidence of this event. If the Universe truly had been compressed into something so dense, there should be an echo of light everywhere. By the 1960s astronomers believed that it could be possible to find leftover radiation from the early Universe. Two scientists were about to do that, but not before having to get their hands dirty.

In 1964, Arno Penzias and Robert Wilson were working for Bell Telephone Laboratories. They were using the Horn Antenna to detect radio waves bounced off Echo balloon satellites: these were balloon-shaped satellites sent into orbit to passively reflect microwave signals, allowing the transmission of radio communications. They had to make several adjustments to the antenna, (even cooling down the receiver with liquid helium) to pick up these transmissions. Unfortunately, they were receiving more than 100 times the noise that the whole device should have produced. They put this down to the discovery of pigeons that had decided to nest in the antenna; surely the birds and their droppings were the source of this noise. Once the birds had been safely trapped and relocated, the pair scrubbed the antenna clean. But the noise persisted; it was consistent, coming from beyond the Earth, the Sun, the entire galaxy. They spoke to colleagues and read up on the subject of possible celestial sources for this noise. That's how they realised they had discovered the cosmic microwave background. I fricking love this story!

'The cosmic microwave background represents the electromagnetic radiation that's travelled the largest distance that we can possibly measure,' Professor Peter Coles (he/him) tells me. Coles is a theoretical cosmologist at Maynooth University in County Kildare, Ireland, an ideal candidate to explain the importance of this discovery and sixty years of observations on our theories of the Universe. He was my first port of call for this. He is not only a world expert on the Universe as a whole, he's also someone who can clear up for me the drama behind the scenes of the cosmology world.

> The detection of the microwave background radiation back in the sixties was one of the things that convinced most cosmologists that there was a Big Bang. We can now use the detailed observations of the microwave background to test very specific components of that theory, like what the Universe was like in the very early stages, what fluctuations were there that could have generated the galaxies, and the large-scale structure that we see now. Its existence is important, but its detailed properties give us even deeper insights.

This is why the CMB is such a revolutionary discovery and remains the number one target in microwave astronomy. It confirmed that the Universe started in the Big Bang and ever since then the CMB has become more than just a simple proof. Observing the CMB is not difficult. A small percentage of the noise in an analogue TV or radio is from the CMB. And such observations will reveal that this background is remarkably uniform. Converted to a temperature, it tells us that the basic temperature of the Universe is 2.726 Kelvin (approximately -270 degrees Celsius). Less than three degrees above absolute zero.

The first map showed this temperature and that the CMB was the same everywhere you looked and in every direction. This is very important. Because the Universe is so big, if we run the clock backwards we end up with regions that were not at all connected in the beginning. So the Universe shouldn't be so uniform. To solve this problem, astronomers have proposed the existence of a brief moment called 'cosmic inflation' that took place before matter even formed. This allowed for the Universe to be a lot more compact in the beginning. There is a lot that we don't know about cosmic inflation, and the CMB will be important in helping us to find out about it, but before we tackle that, we need to realise that even this extremely uniform CMB is not perfect. And we wouldn't be here if it were. Thanks to breakthroughs in observatories, researchers have been able to measure the CMB better and better over the years. It turns out that there are small differences in temperature across the CMB; these are called CMB anisotropies. They are of the order of one part in 10,000.[2] These small differences are important – that's where the galaxies ended up forming.

Now, you might be thinking: 'Alfredo, literally thirty seconds ago you said that everything was smoothed out thanks to cosmic inflation. Now you are saying apart from these small differences that end up being so big that galaxies and clusters of galaxies form in them. What gives?' I know that it seems that the two concepts are almost contradictory but there is something that fixes this (it creates other problems, but a win is a win). The solution is found in quantum mechanics. Basically, reality at the smallest scale is a mess of fluctuations. Even the emptiest vacuum we can create will contain particles and energy. So when the Universe was just a tiny fraction of a second old, there were these minuscule variations, despite being it almost completely uniform. Then *bang* – cosmic inflation blows everything up to enormous proportions, making the Universe expand. Those variations allow for a little bit more matter to end up there when matter forms. Then that extra bit of matter was able to attract more than the rest. So suddenly – well, over millions of years – it was attracting enough matter to make stars, black holes and galaxies, just from one small variation.

The Fault Is Not in Our Stars … or Is It?

As I write this part of the chapter I am in Europe's Spaceport in French Guiana.* I am here to see the first launch of *Ariane 6*, the European Space Agency's latest rocket. It is its predecessor, *Ariane 5*, that has launched many telescopes mentioned in this book. In this section we will talk about the impact of the Planck satellite. I was not involved with it, but the launch was a big deal in my astronomy department. The European Space Agency mission has revolutionised observations of the CMB, providing the most precise map we have ever produced and allowed us to, among other things, work out the expansion rate and the age of the Universe with unprecedented precision. The expansion rate of the Universe is now a major point of discussion and contention in the field. Grab a biscuit, because we are about to get some tea.

For slightly less than a century we have known that the Universe is expanding. We got there thanks to the work of dozens of women, underpaid and underappreciated in their lifetimes, the so-called Harvard computers. They classified hundreds of thousands of stars, producing groundbreaking research and creating standards we continue to use today. Among them, Henrietta Swan Leavitt delivered the insight that changed our

* Wales & French Guiana in a single chapter… this is clearly a well-travelled topic!

understanding of the Universe for ever more. Her discovery of stars known as Cepheid variables. These are pulsating stars whose day-to-month-long changes have an extremely stable period and are closely linked to their true luminosity. In everyday parlance, how luminous and how bright something is are basically the same thing. That's not the case in physics. In scientific terms, luminosity is how bright something really is – we say 'intrinsically'. Think of a 20-watt light bulb. That is its luminosity. But the same light bulb would be brighter if you are in the room where it's on or if you are down the road. The luminosity doesn't change; it's still 20 watts, but its brightness does change because it depends on how far you are from it. In physics, brightness is how bright it looks to an observer. The further away you are, the dimmer the object is. This change follows a very specific law. If you know how bright something appears to be and its real luminosity, you can measure its distance from you. Understanding the relation between pulsation and brightness developed by Leavitt allowed us to use these objects as the first standard candles: a sort of cosmic milestone (in the original meaning of the word) – a way to track the distance. Armed with Leavitt's Law, Edwin Hubble looked for Cepheid variables, in an object then known as the Andromeda nebula. He discovered something that appeared impossible: Andromeda was far beyond the limits of the known Universe.

The year is 1924 and as far as humanity is concerned, our galaxy, the Milky Way, is the entire Universe. In Italy, only one of my grandparents had been born, the one I was named after. In the US, Hubble found twelve Cepheid variables in Andromeda. He used them to estimate that the object was 900,000 light-years away. This is not exactly correct; with more refined estimates we now know that it's 2.537 million light-years away. Still, it showed that this was far beyond our galaxy, which is estimated to be about 100,000 light-years side to side. Andromeda was not a nebula, and the Milky Way was just a galaxy like many others. Suddenly, the Universe was a lot bigger. But Hubble was about to break the Universe again …

He measured the distance of several more galaxies and he also measured their redshift, the concept we mentioned towards the end of the previous chapter. We have said that it is like a Doppler effect, but it is not truly one, because the Universe is expanding. But Hubble didn't know that quite yet, so he was using that approach to measure the velocity at which galaxies are approaching or receding with respect to the Milky Way. Andromeda is actually blueshifted. That's because it is going to collide with the Milky Way in several billion years' time. For all the other major galaxies, Hubble discovered that they are redshifted. They were all moving away from us. This is not what a random distribution of objects moving about should look like. This meant something – the further away they were, the more redshifted they were and the faster they were moving away. The proportional relation between distance and redshift is encapsulated in the Hubble–Lemaître Law, and this proportion is defined by the Hubble constant, the value of the expansion rate of the Universe today. Basically, the speed at which the galaxies appear to be receding from us is equal to the distance times the Hubble constant. So the farther the galaxies are, the faster they appear to move away.

Now, over the last 100 years or so our understanding of the Universe has changed massively. We know that the Universe is not only expanding but it's undergoing an accelerated expansion, the cause of it dubbed 'dark energy', a hypothetical and unknown form of energy that affects the Universe at large scales. The measurement of the Hubble constant has become more and more precise, settling for a specific value, the value we got from the Planck satellite. It served us well until we found something was wrong: this is the debate of the decade in cosmology. Sirs, ladies, and gentletheys, let me introduce to you: the Hubble tension.

The Planck telescope delivered an exquisite map of the cosmic microwave background. It does not measure the Hubble constant of the Universe directly, but you can get to the expansion rate by using a combination of the parameters you can extract from CMB and a model of what happened in the Universe between now and then. While not a direct measurement, there is a lot of data and it is very precise. The value they find is 67.4 kilometres per second per megaparsec (a megaparsec is equivalent to 3.26 million light-years). This means that if a galaxy is one megaparsec away from us, it will appear to be moving away at a speed of 67.4 kilometres per second. For several years this became the gold standard.

Astronomers have not forgotten the technique used by Edwin Hubble, measuring the redshift and distance of galaxies. Cepheid variables are still enormously useful in that quest, but other standard candles have been discovered. One among them is Type Ia supernovae. It turns out that not all supernovae are created equal – some are more equal than others. These events' main protagonist is a white dwarf. A white dwarf is what is left over when a star like our Sun becomes a red giant and then runs out of fuel. Without nuclear fusion to push back against gravity, the star collapses. In more massive stars, this collapse is so intense that it makes the whole star explode. In smaller stars, instead, it is stopped by the atomic forces: there's no explosion, but it still releases a lot of energy, which pushes away the outer layer of the star, revealing its dense, newly exposed core. But just because it's at the end of its life, that doesn't mean it's over. If this white dwarf has a companion, it might just end up doing something incredible. The white dwarf steals material from its companion, and beyond a certain threshold of mass it will collapse into a supernova. And given that the threshold is a specific value, the supernova will always look the same and have the same luminosity. It's pretty nifty of the Universe to provide something like this. Using the stellar distances measurement, astronomers found the expansion rate of the Universe to be around 73.0 kilometres per second per megaparsec (km/s/Mpc).[3]

You do not need a Ph.D. in astrophysics to know that 67.4 is different from 73. Scientists give the numbers of their observations with an uncertainty range called standard deviation (also known as sigmas). So you hear numbers like '19 plus or minus 1', or something like that. This means the best measurement is currently 19 but the true value is expected to be within a few units from 19. The problem here is that these uncertainty ranges do not overlap. The value of the CMB is of plus or minus 0.5 km/s/Mpc, for the stellar method it is 1 km/s/Mpc. The chances of the two being compatible is 0.00006 per cent – which is tiny! When I asked Peter Coles about the tension, he gave me a lot of fun gossip, but he was keen for me to understand this quest first from a historical point of view.

When I started in this field in 1985, I remember the first set of lectures on cosmology I went to. Richard Ellis, a very famous observational cosmologist, gave a talk in which he talked about the controversy over the Hubble constant. There were two groups who claimed different numbers that were discrepant with each other. These were both actually using stellar distance indicators, not the CMB and stellar. One group claimed that it was about 100 megaparsec plus or minus 10. And there was another group who said it was about 50 plus or minus 5. And basically, if you split the difference, you get about 70 but that number in the middle is several sigma away from both of the results. The lesson from history is, basically, people aren't very good at estimating their errors. And without pointing the finger at either of the two current estimates, it could well be that the resolution is that reason again, that basically there's an unknown source of error in one or both of the measurements, but it is interesting. I'm a devout agnostic when it comes to this. I wouldn't like to place bets!

Basically, we can't get two different numbers. The true Hubble constant is one specific value, but the best approach we currently have to estimate gives us two different values. Planck data is not easily redone, although we will see the new observations of the CMB are coming in the next section. Anyway, cosmologists do not seem ready to claim the fault is there.

'The value seems very robust from the point of view of the CMB, but it does require understanding what happens between the very early Universe and now in order to make a value of the Hubble constant, so that extrapolation relies on the model, and there could be ingredients in the model which are not right. That's perfectly possible, and it's quite exciting if it turned out to be true,' Professor Coles explained. In the stellar distances camp, new observations keep coming in, regularly strengthening the case that they are correct. 'They consistently get a higher value than the Hubble constant. And those are actually local measurements, in which they are actually trying to measure the distances of stars and calibrate them, and so on. And those also seem quite robust measurements. They're quite hard measurements to make, but we have lots of, you know, advances in observational technology. Either one of them has got an error in it, or both of them have an error in it, which we don't yet understand … that's possible!'

Recently, a dissenting voice in that camp found a value of 69 kilometres per second per megaparsec, sitting among the two numbers, more than breaching the tension, suggesting that the fault is in systematic errors in our estimates of the stellar distances.[4] But it is not just observational approaches that are being called into question. The Standard Model of cosmology has been extremely successful at explaining so much of what we see, but since it includes dark matter and dark energy and we do not know what they are; there are many theoretical scientists who would happily change it – that's to provide insights into these substances – or to get rid of it altogether. And that is when our conversation with Professor Cole got juicy. He believed, as he said above, that models could indeed require updating and tweaking, but he thinks there is a lot of guff out there.

'The interesting thing is that theorists all the time are inventing stupid models, outrageous models, where things go "decaying of something" or "early varying of something". And it is actually really hard to make any of those models work in the context of the Hubble tension without messing up other things that fit perfectly well,' he tells me, before continuing with a mischievous giggle:

> It's been quite fun to watch the theorists using their ingenuity to try and say: 'Well, I've thought of this new kind of scale,' or 'I feel dark matter is doing this, that and the other thing.' And they write a paper and say, 'We can resolve the Hubble tension,'

> but then the next day somebody comes along and says, 'Yeah, but if you take that model, it doesn't fit these other observations,' which actually the Standard Model does fit very well.

And that is the crux. Similar to the Standard Model of particle physics, the Standard Model of cosmology fits observation well: it has great predictive powers, it has been fundamental to understanding the Universe and yet it has some big unknowns we have yet to solve.

A Wrinkle in Light

Even with the big question mark of the Hubble tension, the work done by hundreds of scientists has become a cornerstone of cosmology. Still, it has been more than a decade and a half since Planck lifted off; we need to consider the future and what we can do next. Dark matter, dark energy, cosmic inflation remain uncertain ideas. The latter in particular might be solved by a feature in the CMB that will require new and more precise observations. Inflation helps us solve the massive paradox that we see in the uniformity of the Universe, but let's not pretend it isn't wild. No known particles or forces can make inflation happen: an inflationary field is responsible for it. I know, I might have just as well have told you that unicorns are responsible for it. I completely agree that these concepts are very nebulous if not completely wishy-washy, but there is maths behind them. The problem with the inflationary field is that it is also nebulous, for cosmologists. We have multiple ways of making it happen and we do not know which one is right. To find the right one we need to look at signatures of this event frozen in the light of the CMB. We need to measure the polarisation of the CMB.

In everyday life there are two places where you encounter the polarisation of light as a specific concept: your sunglasses and 3D glasses at the cinema. Light is electromagnetic waves, so you can imagine the electrical components going up and down on a plane perpendicular to the plane where the magnetic component is. Very basically, if you are looking down at this wave it might look like a cross +. The electrical part goes up and down vertically, the magnetic one horizontally. But the cross can rotate. It can look like an x or any direction. If all the light from a source looks like +, then the light is polarised. Polarised sunglasses only let light with a certain polarisation pass. Since sunlight is unpolarised, this cuts off a lot of sunlight. In 3D cinemas, the film in question has two images with different polarisations being projected, but the lenses only let one get to a specific eye, blocking the other.

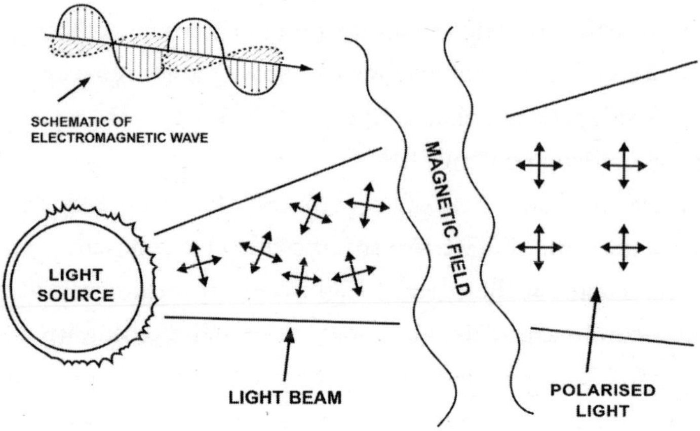

A schematic view on how magnetic fields in space can polarise light.

Credit: Alfredo & Chris Carpineti

The CMB is also polarised. Unfortunately, we are not going to see it with 3D glasses. Although, how funny it would be if there was just a message there or a figure like those 3D illusion pictures you need to cross your eyes to see. If there was one, my money is on Douglas Adams's Last Message From God to His Creation: '*We apologise for the inconvenience.*' The CMB would be a pretty great place to leave a message, since it is seen all over the Universe. Plenty of scientists and non-scientists have considered it as a place to find answers to what existed before the Universe, and not all these signs are to find evidence of deities. One suggestion: the cyclical Universe hypothesis assumes that the Universe explodes out, expands and then contracts back on itself before collapsing. Then it explodes out in a new Big Bang and the cycle starts anew. It's called the Big Bounce. Polarisation in the CMB is not to find what came before the Big Bang, though, but it is to find what happened right after it.

'Inflation took a singularity-sized Universe to a cosmological-scale Universe in fractions of a second, 10^{-32} seconds. In that process, the field that drove this rapid expansion would produce fluctuations in the gravitational metric. So a spectrum of gravitational waves,' Dr Max Silva-Feaver (he/him), a postdoc at Yale, walked me through the truly first instants following the Big Bang.

> The gravitational waves would just permeate the Universe and continue radiating, and in time would interact with the light and the particles you get as the Universe cools. The CMB is eventually released 380,000 years after, and those early gravitational waves would leave a particular polarisation pattern on the CMB. If we make a map of the polarisation of the CMB, we would actually expect to see these curls, where the gravitational waves were.

Making this map is the near-future goal for astronomers working on the cosmic microwave background. Dr Silva-Feaver is working on the analysis pipeline of one of the telescopes at the Simons Observatory currently under construction in the Atacama Desert in northern Chile. His work is important in assessing the capabilities of the telescope and all the data necessary to create the future map. Silva-Feaver was also involved in the development of the sensor package, as well as the software to get the data out. Thanks to his expertise we can get to the challenges of observing the CMB from the ground and how, in a matter of years, we might be getting exciting insights into the polarisation of the CMB, but not just that. There are four telescopes in the observatory. Three are small, just forty centimetres in diameter and one is big, six metres across. The four telescopes have slightly separate goals, each worthy in its own right. As is often the case, the total will be greater than the sum of its parts. The large telescope is doing a lot of work. It is designed to collect microwave and submillimetre observations that will be used to better understand how galaxy clustering and dark matter affected the light of the CMB across the ages of the Universe. It will provide constraints on neutrinos, even providing new understanding into their elusive mass. It will track the behaviour of dark energy. It will improve the parameters used to estimate the Hubble Tension. Truly, it already sounds like Christmas has come early for cosmologists.

The work of the smaller telescopes will also be important. They won't be looking at the whole sky but a portion of it, about 10 per cent. It will be a smaller but deeper search. Here they hope to see the polarisation of the CMB and evidence of the gravitational waves created by cosmic inflation. The telescopes are small but a feat of engineering nonetheless. They have been built in a high and dry place (the Atacama Desert) to reduce the effect of the atmosphere. Part of the instrument will be kept at very low temperatures, close to absolute zero, with a sapphire half-plate rotating in front of the mirror. This rapid rotation causes a modulation in the light that comes in, as long as that light is polarised. A massive challenge, I am told. The other major challenge is that the CMB, as we shall see further in this chapter, is not the only signal in microwave and submillimetre astronomy out there. There's very cold dust in the Universe that releases microwaves and submillimetre light. To see the signal from the CMB, that needs to be carefully removed. The signal would be confusing at first, but, taking a rainbow of microwave observations, you can separate them, too.

We have to actually build telescopes with many different spectral bands, many different kinds of filters, essentially. We have some telescopes that are just dedicated to measuring the dust so that we can make a measurement of the sky dominated by the dust emission and a measurement of the sky dominated by a combination of the CMB and the dust. And then we can use the two telescopes to sort of subtract away the dust contribution, knowing that they have different spectral dependencies. If we didn't have dust we wouldn't have to build so many telescopes, Dr Silva-Feaver explained.

Saving money by having fewer telescopes would be good. But without that dust we might not have seen so many other phenomena in the Universe. And we are about to see that, without this 'annoying' dust, we wouldn't be here at all.

One Scientist's Problem Is Another Scientist's Treasure

We began this chapter with Catholicism, met some scientific dogmatism along the way, and now we are back to Catholicism. 'Remember you are dust, and to dust you shall return.' We could make ourselves feel better by defining ourselves as being made of stardust, but this is not exactly correct. Don't get me wrong: we are mostly made up of elements that were forged in the heart of stars plus a bit of hydrogen from the Big Bang. But stardust is a smaller fraction of the cosmic dust that ends up being the building blocks of humans, capybaras and whole planets.[5] It's not a fraction that we can discount, though. At the same time, cosmic dust is more than a nuisance for CMB observations. It opens a whole new window into stellar, galactic and even cosmic properties. In the last chapter we saw that we can use radio waves to work out the presence of molecules in space. That's how we know that ethyl formate – the rum/raspberry molecule – is present near the centre of the Milky Way. Shifting to slightly shorter wavelengths, we get the emission of other molecules, the most interesting being water and carbon monoxide. Stars are the cause of these excitations, especially young bright stars, which are born in clouds of gas. Their ultraviolet light is absorbed by those clouds of gas and dust and then the light is emitted in infrared and longer wavelengths. To understand how to use this emission to study the Universe, I spoke to Dr Yang Chentao (he/him), now a software developer but previously an expert into this cosmic dust. He tells me that his first name, Chentao, means 'star waves'.

Dr Yang began by explaining to me the importance of doing these kinds of observations to understand both how stars form and how they shape galaxies:

> I use these molecules, this emission, like a tool to study what the temperature of the gas clouds is, what the density is, and how they move in galaxies. This gas is probably located where you have a lot of star formation going on, where a lot of young stars are created. So by probing the property of this gas, maybe I can have a clue into how star formation is happening in these galaxies. Also you can learn something about black holes as well.

Before we get to black holes, though, let's stress the importance of the other two focus subjects. The formation of galaxies is uncertain. We have a rough idea of how they evolve, but we have massive gaps in knowledge. And let's not get too much into stars, but there is also a lot of vague, hand-wavy: '*they sort of form like this*' and '*they change like that*' and then '*they die like this*'. Ignorance is a fantastic motivator for humanity. We want to make sense of the Universe. Microwave, millimetre and submillimetre astronomy, the loosely defined focus of this chapter, provides some important insights into the big questions surrounding star formation and galaxy evolution. These wavelengths can actually help astronomers find star-forming regions in the first place. Once you have them, astronomers can study the polarisation of light in stellar nurseries, where stars are born, which provides maps of the magnetic field of these regions. This is a fantastic new exploratory dimension for understanding these structures. It is easier to track the gravity of these systems, but we can't dismiss the importance of magnetism, as it comes into play in the most dramatic stellar processes such as pulsars, FRBs and more.

When I think of these observations, I always feel like they are the L-shaped block in Tetris. Often they won't solve your whole game, but they are very good for your long-term strategy. From the renewed comprehension we get into stellar nurseries to better understanding the evolution of galaxies, it's a short step. Galaxies are enormous compared to most stellar nurseries, but the distribution of gas and dust across these so-called 'island Universes' depends on the stars that form and die, and how their light and winds of charged particles sculpt and push away the leftover gases. The microwave light we are looking at here puts the stellar nurseries into the galactic context, even in galaxies that are pretty far away, where the light of these regions has shifted to longer wavelengths. These wavelengths are so important, also because celestial objects can create an emission similar to lasers.

Laser stands for 'light amplification by stimulated emission of radiation'. Simply put, a laser creates light of a single colour (a specific wavelength/frequency) thanks to quantum mechanics. Light passes through a medium (a gas or crystal), exciting its atoms. That means the electrons have a bit more energy and they are on a higher orbital. Not enough energy to escape the atoms, as in the epoch of reionisation in the early Universe, but a bit of extra energy. The atoms will lose that energy and emit light of a specific wavelength. Suddenly you have a nice monochromatic light. The same principle is behind the astrophysical masers and, as the name suggests, the light in question is in microwaves. Masers are found in many different environments. The aurorae on Jupiter that we will encounter in the 'Ultraviolet' and 'X-rays' chapters have a maser component. Masers are found around comets. They are found in the atmosphere of old stars. They are found in the stellar nurseries. And, of course, they are found around some supermassive black holes, the black holes that we believe sit at the centre of almost every galaxy. Some of these black holes are surrounded by a disc of material. Maser lines are emitted by water molecules. Like a laser, they are a specific colour, which means a specific wavelength or frequency.

Dr Yang continued:

> People use this 22-gigahertz water line to trace the disc surrounding a supermassive black hole. You can use this to figure out the kinematics of the disc. And they also use high-resolution observations to map the structure. And using this tool you can measure the black hole mass and also the size of the accretion disc and maybe even the distance. If you derive the geometric size of the disc, you can measure the distance independently from all other methods.

I love how important these observations can be for black holes, but I am going to further stress the part about the distance. This method allows you to work out the physical size of this disc. You can measure how big the disc appears in the sky. Using basic geometry, you can measure the black hole's distance from us. And, as we learn in this chapter, we really need some good measurement of distance for galaxies. Yes, a major application of this is different estimates of the expansion rate of the Universe. So why don't we have a measurement of the Hubble constant with the masers of supermassive black holes? We have,[6] but while it is closer to the stellar measurement than the CMB, the uncertainty is such that we can't solve the tension with it. Well, not yet. There are challenges to studying these wavelengths, making precise maps and getting the high-quality information out. Still, astronomers are working on them, looking at the detailed events around supermassive black holes and opening the door to understanding the whole Universe.

And on the Third Day …

The chapter would not be complete without the revolution that has been the observations from the Atacama Large Millimeter/submillimeter Array (ALMA), located in the Chilean desert. ALMA uses the part of the microwave range closest to infrared where wavelengths are around a millimetre. Just over a decade ago, the observatory began its high-resolution mode for observing young stellar systems. That's truly when everything changed for this particular field. ALMA observed a young star called HL Tauri. One million years old and some 450 light-years from Earth, this is a Sun-like star, truly in its infancy. Around it, there is a disc, a protoplanetary disc, from which planets will form. But the exquisite resolution of ALMA provided more than just that. It showed that the disc of dust and gas, left over from the star-forming process, is made of concentric rings. In each groove, astronomers believe a planet is forming, creating a gap in the disc. Those observations alone sell us the whole field of millimetre astronomy, and since then we've had so many more. Put simply: if we can't 'see' it, we can't study it. What this wavelength range delivers for protoplanetary discs cannot be done in other wavelengths. HL Tauri in infrared or visible light is 'optically thick' – basically, you can't see through the dust. But you go to lower wavelengths and suddenly this disc is perfectly visible. There is nothing obscuring your view; you are getting a front-row seat onto the central portion of the disc, where planets are being born. Dr Wladimir Lyra (he/him), an astrophysicist at New Mexico State University in Las Crucas, is using those views to understand how planets are born. That's why these observations are so important.

'To see the middle of the disc, you need to see where the disc is transparent. It is not transparent in optical. It is not transparent in infrared,' Dr Lyra explained.

Millimetre astronomy provides the means of seeing the process of planet formation. It does so by showing the effects of the growing planets on the disc, such as the grooves of HL Tauri. It reveals that in cases such as protoplanetary system HD 163296 the gas and dust move around in it like enormous waterfalls.[7] It can also show the planet directly, like the incredible system featuring planet PDS 70b which appears to have a second planet forming in the same orbit – the *first ever* evidence of a co-orbital planet.[8] What a find! Still, all these incredible systems don't provide all the answers to the questions of planet formation, but they help a lot.

'It's kind of a tough question to understand how planets come to be. We need to use several different tools. We need to use the evidence from the solar system, the evidence from exoplanets, the evidence from protoplanetary discs, and models. Putting pen to paper, plus computer simulations go a long way,' Dr Lyra continued.

We think that planets, no matter how big or small, started from dust grains, the same dust grains that annoy CMB observers and that are so useful for tracking star-forming regions. They are, in fact, found around stars. They are thinner than a human hair, but these protoplanetary discs are big doughnuts of gas and dust, and these grains interact, collide, merge and grow. Suddenly you have a pebble. And a pebble has more chances to interact, collide, merge and grow under gravity and other forces at play. Suddenly, it's growing bigger and bigger. Before long (i.e. a few million years), you will get a planet.

But it might not be the only way to get a planet. Brown dwarfs are a type of stellar object that formed like a star, from a collapsing gas cloud, but were never big enough to start nuclear fusion. Some aren't much bigger than Jupiter. Could gas giant planets instead form like stars do? Maybe. It's an open question in our understanding of the processes at work, and it is not the only one. Why do we have terrestrial planets and gas giants? Why do certain gas giants orbit so close to their stars? How common are systems like our own? How common are planets like Earth? I think we have a solid sketch of the formation of planets, but it's not complete and we can easily change and expand what we currently have.

Dr Lyra told me about the challenges in making the models. There are several uncertainties that involve guesswork. It is difficult to balance the need for simplicity in computer models with getting something out that we can understand and match reality. More observations are always needed. In particular, the details we got from ALMA in the last decade are still unmatched, especially in longer wavelengths (such as radio waves). These will allow more insights in the smaller worlds forming in these discs. Not just the Jupiter-sized planets but also the Neptune-sized. Maybe even ones as small as Earth-sized worlds. This won't be the only approach to solving the mysteries mentioned and the many more open questions, but without the details brought in from these very early moments in the history of a planetary system we cannot expect to create a perfectly correct model. Just because you can see a pretty cake doesn't mean you can tell all the ingredients needed to make it without further investigation. Millimetre astronomy is a snapshot of how the cake is being made, and a little bit about the ingredients that go into it; not enough to actually make the whole cake, but we can agree it would be much harder to do without.

In our journey beyond the visible we have jumped from the radio waves produced by so many different objects to microwaves which are emitted by some of the coldest objects in the Universe. Certainly, there's the cosmic microwave background; you don't get colder than that but the emission of planet-forming dust around so many interesting places. Our quest to see the invisible has made us to ponder the greatest mysteries of the Universe. The question '*how are we here?*' is written on a cocktail napkin and the ink has bled all over the place. We were convinced we had the answer last night, but this morning only 50 per cent of it makes sense. We know of the Big Bang, of star formation, of planets coming into being from cosmic dust. The big picture is right there, pretty clear, and we can be fairly confident about that. The details are filling in the gaps and there are a lot of gaps to fill. We do not shy away from this work. What's more fun than discovering if we are right or wrong?

— CHAPTER 3 —

INFRARED

The Universe at the Edge of Our Senses

And now here is my secret,
a very simple secret:
It is only with the heart that one can see rightly;
what is essential is invisible to the eye.
—Antoine de Saint-Exupéry

AWAKE! for Morning in the Bowl of light
Has flung the Stone that puts the Stars to Flight,
And Lo! the Hunter of the East has caught
The Sultan's Turret in a Noose of Light.
—The Rubaiyat of Omar Khayyam
(Edward Fitzgerald)

Of all the wavelengths of light beyond what our eyes can see, infrared is the one we have a physical experience of. Feeling the heat coming from something – a sense known as thermosensation – is just our body 'observing' infrared light. So even if we don't know that it is another form of light, we know that there is something there. And we have known it since birth: the first light we feel is the warmth of the bodies in the delivery room and so infrared light is intrinsically familiar.

The first time I remember coming across infrared light (though

I did not yet know what it was) was while playing with the TV remote in my grandparents' place. They had one of those enormous rubber protectors which were so common in the nineties. To a child a remote is essentially magic, and I guess I wanted to play sorcerer's apprentice because I remember experimenting by putting several materials in the little opening on top of the protector to see what would work. What I was doing (without knowing it) was testing what material was transparent in infrared. I remember vividly one test that puzzled me a fair bit: it did not seem to work through the frosted glass door leading to the living room. One wavelength's transparency might be another wavelength's opacity.

For others, the most common memory of their first appearance of infrared will be from movies and TV – in night-vision goggles, usually, but also through supernatural shows where the fictional ghostly presences trigger a slight change in temperature. My first encounter with that was in my favourite movie. Who can forget Tim in *Jurassic Park* using night-vision goggles before before Lex wonders what happened to the goat!

Whatever your first experience with infrared light, there is something common with all of them: the experience of learning about an unseen world, making the invisible visible, and uncovering what has been shrouded from our senses. And all this is just too much of a suitable metaphor for how astronomers use infrared light and light beyond the visible in general. We can divide the use of non-visible light into two big categories: using it to see new objects or new properties of known objects and using it to discover what those objects are made of. This classification is obviously exceptionally broad, but it is important to keep that in mind as we will find how astronomical investigation boils

Sunlight reflecting on the Kraken Mare, a northern sea of Titan in this infrared image.

NASA/JPL-Caltech/University of Arizona/University of Idaho

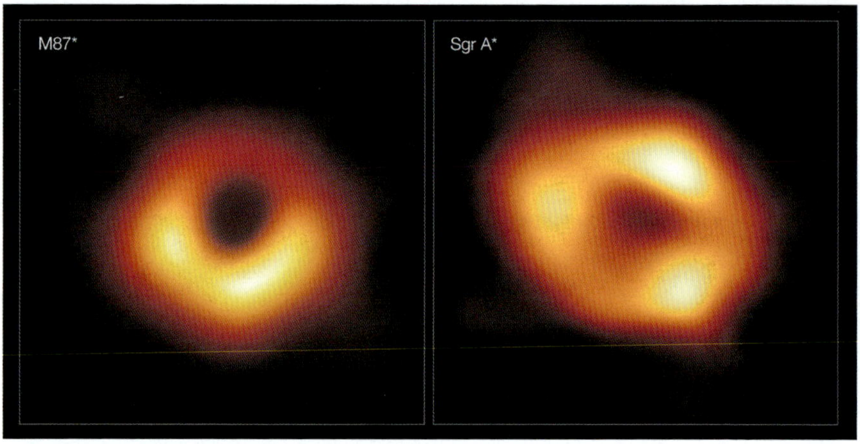

The extraordinary images of the shadows of two black holes: M87*, which is 55 million light-years away, and Sagittarius A*, the supermassive black hole at the centre of our galaxy. M87* is more than 1,000 times bigger than Sagittarius A*.

EHT collaboration

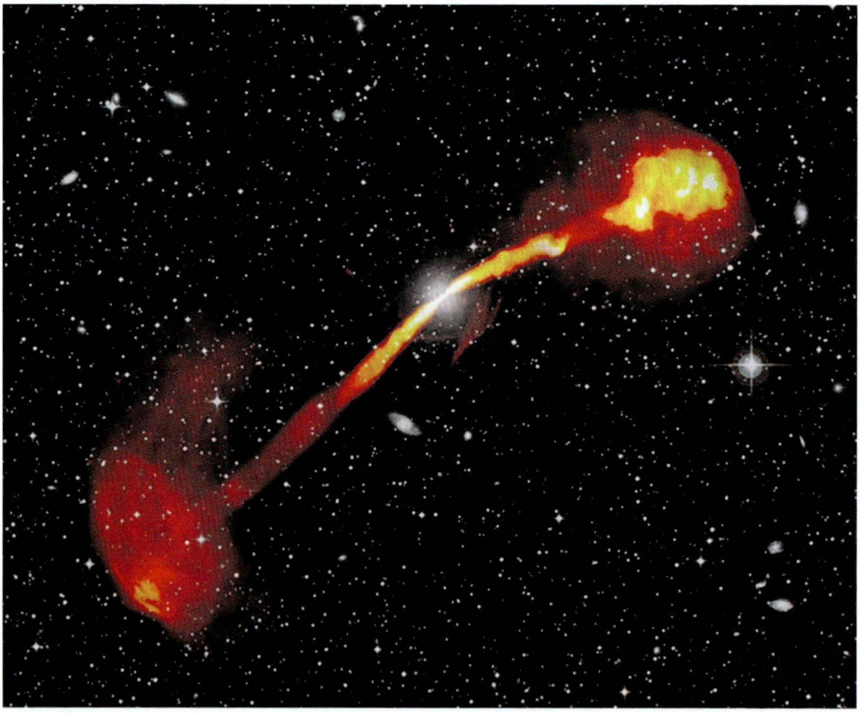

Radio galaxy IC 4296 has an emission that makes it ten times as wide as the Milky Way (in red). It is 160 million light-years but it would be larger than the full Moon if we could see radio waves with the naked eye.

SARAO, SSS, S. Dagnello and W. Cotton (NRAO/AUI/NSF)

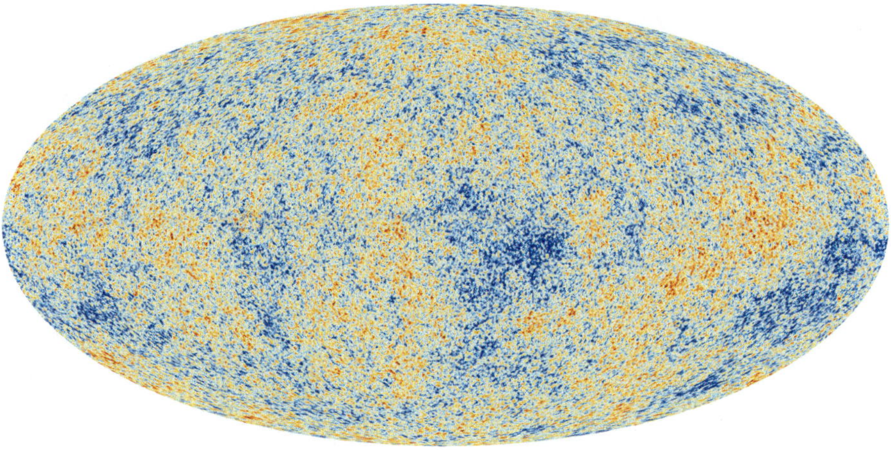

The cosmic microwave background as seen by the Planck mission, highlighting the anisotropies in temperature.

ESA and the Planck Collaboration

At the time, the sharpest image ever taken by ALMA; it shows a baby planetary system. The gaps in the disk indicate where a planet is being formed.

ALMA (ESO/NAOJ/NRAO)

The arc is a gravitationally lensed distant galaxy, and a unique alignment allowed researchers to see Earendel, the dot spotted across the arc.

NASA, ESA, CSA, STScI/Dan Coe (STScI/AURA for ESA, JHU), Brian Welch (NASA-GSFC, UMD)/Zolt G. Levay

More than 45,000 galaxies are visible in this extraordinary deep observation by JWST of the GOODS-South field. Hundreds of them had never been seen before by humans.

NASA, ESA, CSA, Brant Robertson (UC Santa Cruz), Ben Johnson (CfA), Sandro Tacchella (Cambridge), Marcia Rieke (University of Arizona), Daniel Eisenstein (CfA)

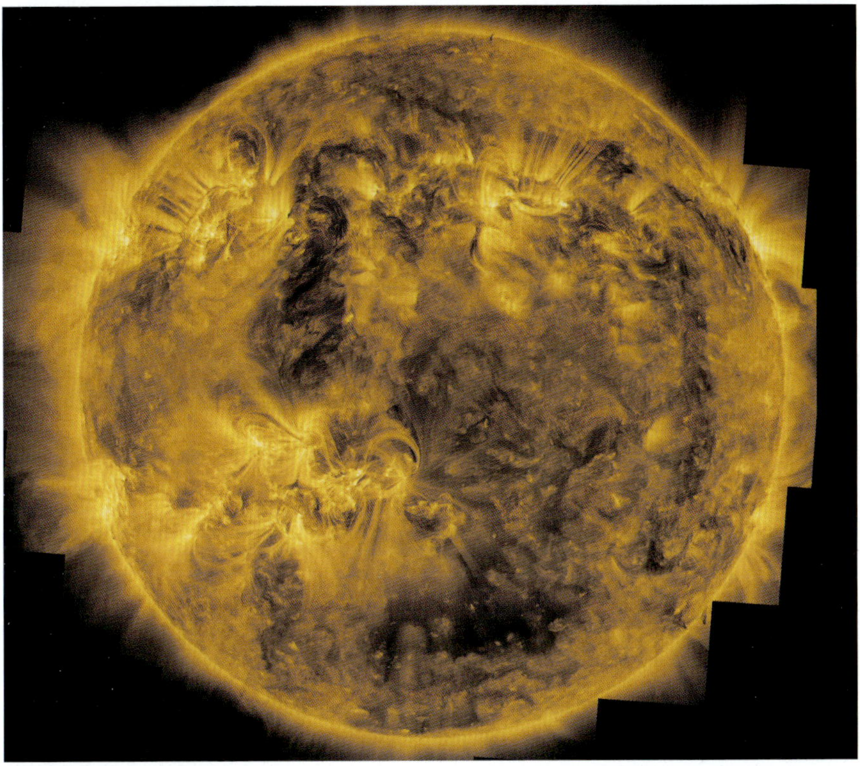

High-resolution image shows the Sun in ultraviolet light taken by Solar Orbiter on 22 March 2023.

ESA & NASA/Solar Orbiter/EUI Team

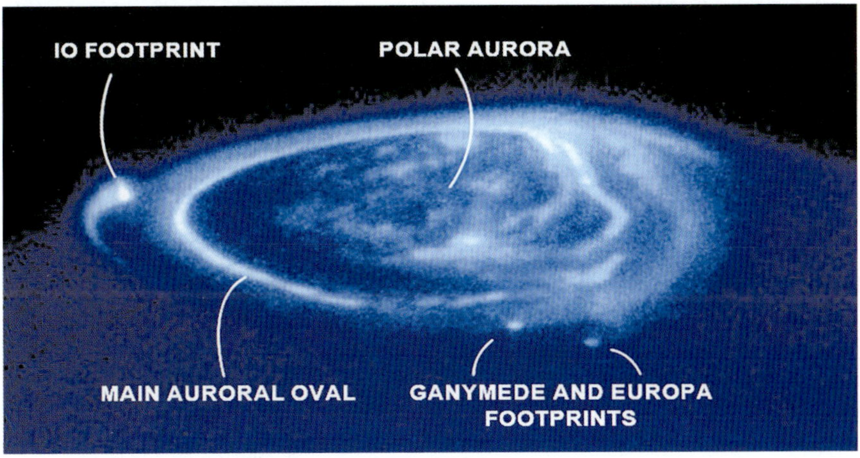

The Hubble UV observation of Jupiter showing its peculiar aurorae.

NASA/ESA, John Clarke (University of Michigan) with modification by Alfredo & Chris Carpineti

The Andromeda galaxy seen by NASA's Swift satellite. This is the highest-resolution view in UV of this galaxy.

NASA/Swift/Stefan Immler (GSFC) and Erin Grand (UMCP)

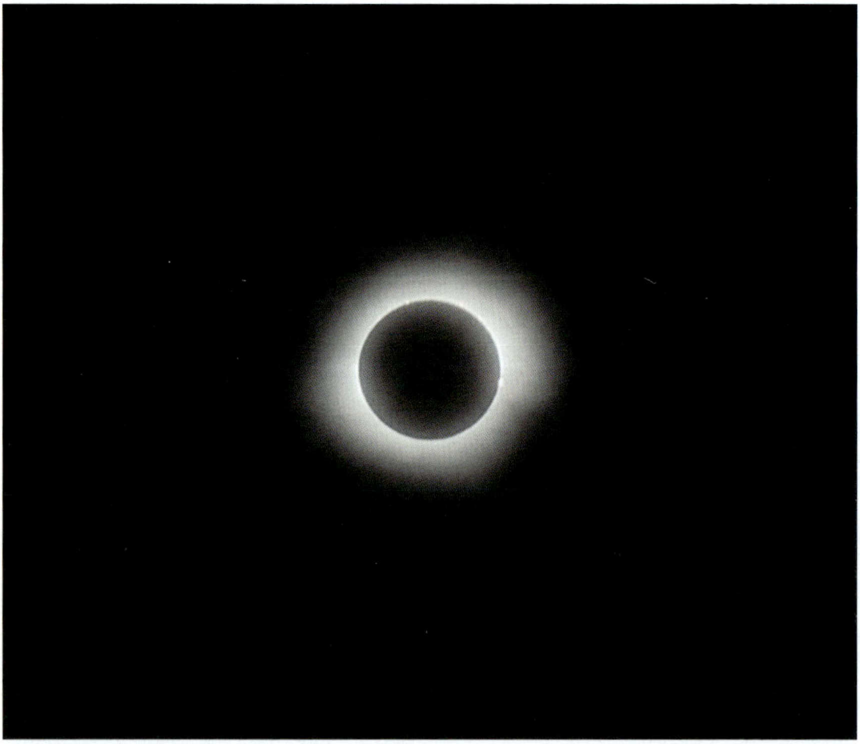

A snap of the 8 April 2024 solar eclipse as seen from Mazatlán in Mexico.

Alfredo & Chris Carpineti

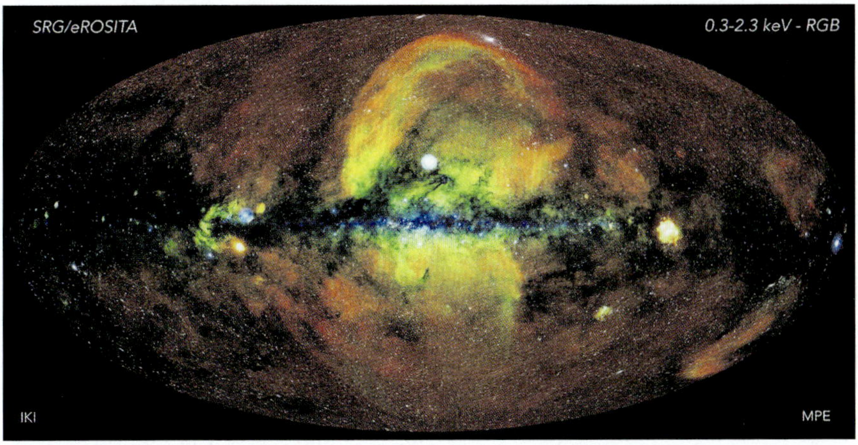

The energetic Universe as seen by the eROSITA X-ray telescope.

Jeremy Sanders, Hermann Brunner and the eSASS team (MPE); Eugene Churazov, Marat Gilfanov (on behalf of IKI)

A million-second observation of the Whirlpool galaxy in X-rays. Most sources (seen in purple) are X-ray binaries.

NASA/CXC/Wesleyan Univ./R. Kilgard, et al; Optical: NASA/STScI

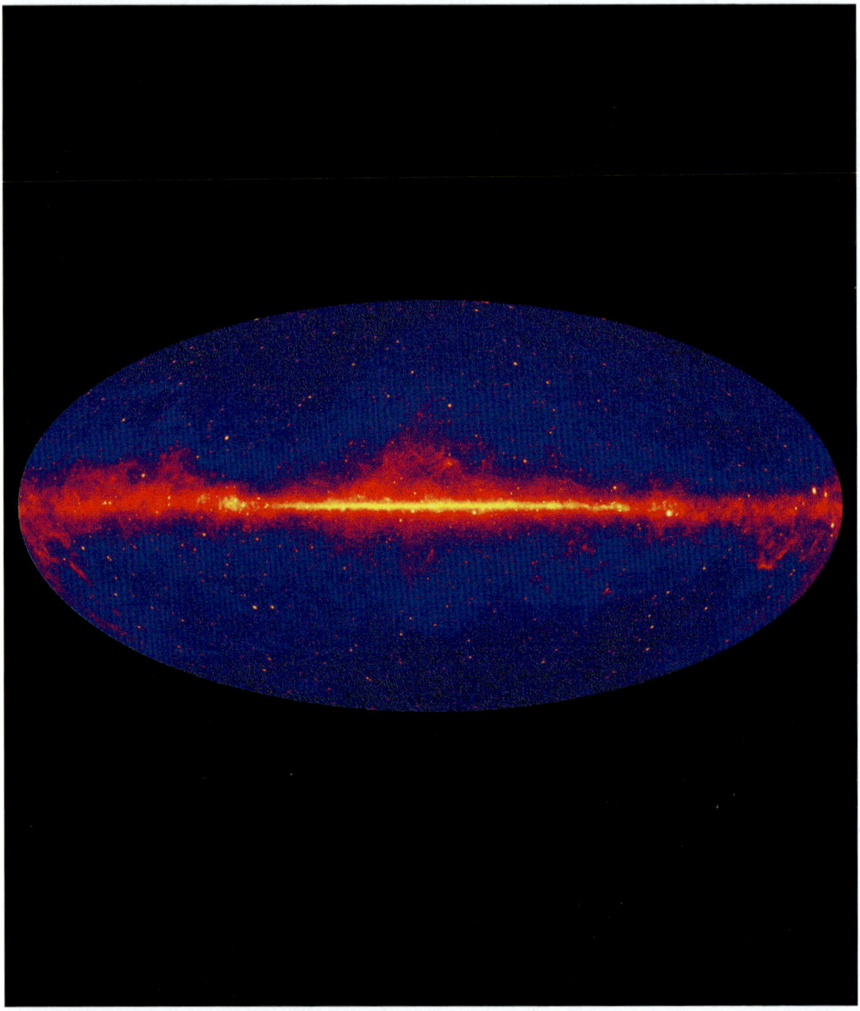

It took twelve years of observations to create this map of the sky in gamma rays.

NASA/DOE/Fermi LAT Collaboration

down to either/or both time and time again. A classic example of something invisible to our visible telescope but clearly seen in infrared is the rings of Uranus, where they are as clear as day.

The Pale Red Dots

There is another exciting thing that infrared light allows you to see that other wavelengths don't: very distant galaxies. The reason infrared light is especially good for this has nothing to do with the light, per se. As I mentioned earlier, the Universe is expanding and that expansion leads to the phenomenon known as redshift. When studying galaxies there are two reasons why this is important. First of all, apart from a handful of nearby galaxies, all others are redshifted. They are all apparently moving away from us. I say 'apparently' for that little caveat discussed in the 'Radio Waves' chapter: they are not exactly moving away from us, so it's not a real Doppler shift. It's the expansion of the Universe that makes it look that way. Second, the light of a distant galaxy will appear redder than a galaxy closer to us, so the more distant a galaxy is, the more its light is pushed into the infrared. And that is not all.

Looking far away into the Universe is like looking back into the past because it takes time for light to travel from an object to you. The further you look back, the more redshift light takes. So there will be a point where you no longer see an object with visible light. There are galaxies that would be visible if they were here, but they are so far away that you need infrared to see them. While infrared becomes key, you don't throw away visible light. The insights we have from these nearer galaxies, those at low redshift, come from the light we can see. But when we translate that into infrared and we look for certain characteristics, we end up with a new understanding of the one farther away at high redshift. Infrared opens a door to the most distant galaxies we might ever see, and we need a guide for that trip. Professor Mike Gladders (he/him) was the person for the job. He has carried out work on almost all the large telescopes in existence, so he has all the know-how on the big observatories, something we discussed in our chat when he told me about his life and work. It was an eye-opener to hear his journey all the way to becoming a professor at the University of Chicago. In particular, the importance of infrared astronomy and how the synergy between the different types of astronomy is so crucial.

'If you want to study distant galaxies, ultimately you rely in great part on an understanding of stellar populations, understanding of nebulae, gas, dust that has been developed over decades by studying nearby galaxies which are many magnitudes brighter, much finer spatial resolution, greater angular scale on the sky. And those low-redshift observations anchor so much of how we understand the distant cosmos,' Professor Gladders explained. The big and powerful telescopes are key but there is a natural process that helps us to see the very distant, and it is Professor Gladders' true bread and butter. I'm talking about gravitational lensing.

Gravitational lensing is another peculiar feature of our cosmos. Space-time is warped by mass. Einstein taught us that and we have seen in the 'Radio Waves' chapter that this has the potential to become extreme, as in the case of a supermassive black hole and its photon ring. You do not have to be a black hole to warp space-time. You and I do it, too, but not in a way we can appreciate it. In space, as long as an object is beyond a certain density, it warps space-time enough to make it act like a lens: it will distort the light of an object in the background and even magnify it. Massive structures are like natural telescopes that allow us to see even further than our regular telescopes will allow. This is what enabled Hubble to spot the most distant star system, called Earendel,[1] whose light travelled for almost 13 billion years before it reached us. While that is quite extraordinary, it is only one example among many. Gravitational lensing is very useful. Professor Gladders used two major infrared space telescopes, NASA's Spitzer and ESA's Herschel (both retired now), to see details of galaxies and to measure the distribution of dust intruding on all new star formations.

'You wouldn't have been able to do this on this object if you didn't have gravitational lensing magnifying the object and hence making it larger in the sky and also brighter overall,' he excitedly explained.

Studying these details is so important. We have mentioned how the infrared studies of these distant galaxies are based on our understanding of nearer galaxies, but the distant ones are not like carbon copies of galaxies today. What they may have looked like billions of years ago might be radically different from a local structure. It is mostly in the last two decades, and perhaps even more recently, that the pieces of the jigsaw have begun falling into place. We are now seeing how galaxies have evolved over the ages of the Universe. For a complete view, we need to go into even more detail and even further back. Luckily, we recently got the right infrared telescope for the job: JWST.

The Biggest Eye on the Cosmos

JWST launched on 25 December 2021, and I hijacked Christmas lunch to watch this *Christmas launch* on the huge projector screen my Italian family usually reserves for football matches. This revolutionary telescope's incredible resolution allows us to see further and more clearly than ever before. A galaxy discovered by Hubble that previously looked like a splotch was determined by JWST to be one of the most distant and oldest barred-spiral galaxies, whose light comes to us from 11 billion years ago, challenging some of the expectations of galaxy evolution.[2] Our galaxy is a barred spiral, which means that the central, older group of stars is not in a neat sphere surrounded by spiral arms, but has a more cylindrical configuration. And, well, astronomers didn't think there would be barred spirals so early in the history of the Universe. But that is not all. JWST has even seen detailed gas composition around supermassive black holes of a nearer galaxy, like no other telescope before.[3] The record for the most distant galaxies ever spotted has been broken multiple times by JWST,[4] objects truly among the first that appeared in the Universe just a few hundred million years after the Big Bang. Understanding these galaxies, their stars, and their compositions underpins our whole paradigm of galaxy evolution. The light of these galaxies is dramatically redshifted, as we have learned.

'I am trying to find that first light we would see from these really distant galaxies. Because of how far away they are, that first light is actually in the infrared, so it's incredibly hard to find these sources,' Dr Taylor Hutchison (she/her) told me. Dr Hutchison is a postdoctoral fellow at NASA Goddard, and our conversation was electrifying. She has a vast array of questions she hopes to answer with JWST about these first galaxies, and, impatient person that I am, I want an answer to all of them so quickly! 'What kind of stars are in these galaxies? What elements are in them? Is it just hydrogen and helium like we would expect in a really early Universe – or do we have some chemical evolution happening? How many more heavy elements are you seeing? How old are these stars? You know, what kind of galactic evolution is happening in this really early Universe?'

By bringing much-needed details to these distant sources, astronomers can compare how galaxies looked and behaved over most of the Universe's 13.8 billion years. It has made the task of comparing faraway objects akin to comparing an apple you are holding in your hand with another apple, rather than comparing an apple with a distant smudge that you think might be an apple.

'The fact that we can now really compare the light we're seeing from these really distant sources to galaxies a little bit closer to us, that helps us have a much better understanding of how galaxy evolution actually happens,' Dr Hutchison continued. 'I'm also excited for whatever weird things we find with JWST that we're not expecting. What kind of weird physics is going to break something that was in one of our theories about how galaxies form and evolve?' She laughed. 'I'm waiting for it because it's going to happen. Because it happens every time you shoot a telescope up in space!'

So here's to the unexpected. After all, the most important phrase in science is not 'Eureka', but 'hmm ... that's weird' or even better, 'what is that??!'. I just cannot wait for what JWST is going to discover, and we will discuss some more incredible observations from this telescope later in the chapter. But before we do that, we need to take a painful but necessary detour. As you may have noticed, I have not yet said what JWST stands for. An acronym usually comes with an explanation, but I have no intention to giving one now, or ever, as the following text will make clear.

Something About Skeletons and Closets

The observatory has been named after James Webb, a former NASA administrator during the *Mercury*, *Gemini* and some of the *Apollo* years, who played a crucial role in what became known as the 'Lavender Scare'. Similar to Senator Joseph McCarthy's better-known Communist witch-hunt of the 1950s, the 'Red Scare', the Lavender Scare was the state-sponsored anti-LGBTQIA+ witch-hunt that purged gay and bisexual scientists and civil servants[5] from US federal jobs in the 1950s and 1960s. Among the people fired because they were presumed to be gay were NASA employee Cliff Norton and astronomer Frank Kameny; Kameny's legal battle against the unfair treatment he received is remembered every year on 18 November, the International Day of LGBTQIA+ People in STEM. This is an initiative that I was personally involved in creating on behalf of my charity Pride in STEM. The discrimination that LGBTQIA+ people in astronomy face is not ancient history, and the effect of those policies remains today. For example, it was only in June 2020 that the US Supreme Court made it illegal to fire someone for being LGBTQIA+ across the whole country.

JWST, which was in development for 25 years from 1996 and

was originally called the Next Generation Space Telescope, was renamed after Webb without any consultation with the astronomy community by NASA administrator (2001–4) Sean O'Keefe in 2002. Webb's role in the Lavender Scare was not known at the time and it only came to light in 2015. Archival evidence showed that Webb facilitated and supported homophobic policy discussions[6] among US senators as an undersecretary of state and he was NASA administrator when Norton was fired in 1963. Since 2002, some astronomers have grumbled about the unorthodox decision to name the telescope after an administrator without community consultation, but there was no major push to change the name. Things did change following the publication of an opinion piece authored by cosmologist Dr Chanda Prescod-Weinstein, astrophysicist Dr Sarah Tuttle, astronomer Dr Lucianne Walkowicz and astrophysicist Dr Brian Nord, published in *Scientific American*,[7] which called for NASA to rename the telescope. This group also started a petition signed by thousands of astronomers. This sentiment has been echoed by other prominent astronomers and science communicators, including Phil Plait,[8] Holden Thorpe,[9] Matthew Francis[10] and Jason Wright.[11]

This is, unfortunately, a moment where you realise how sausages are made. First, let's look at how this metaphorical sausage was presented. Before the launch, NASA began an investigation into Webb's policies and actions while administrator. NASA and its administrator promised a report, but before one materialised in 2021, NASA administrator Bill Nelson claimed that there was no evidence that warranted changing the name JWST. In October 2021, many people and journalists were asking for the report, myself included.[12] NASA replied that there was nothing to publish because the investigation found nothing. This was not good. First, nobody in the academic world would accept a student saying that they did not do some coursework because the answer was zero or the hypothesis wrong. Second, the absence of a report did nothing to exonerate Webb. It actually frayed the relationship with LGBTQIA+ people and allies in astronomy more. But, hey, NASA decided to go double or nothing. A report conducted by Brian Odom was eventually published on 18 November 2022, which stated the following: 'In conclusion, to date, no available evidence directly links Webb to any actions or follow-up related to the firing of individuals for their sexual orientation.'[13]

We know what went into these particularly nasty sausages thanks to a Freedom of Information (FOI) Act request by the journal *Nature*:[14] NASA has found a lot more than they claimed about Webb. A message sent on 3 September 2021, by a person whose name has been redacted, to Brian Odom and Catherine Baldwin explains over multiple, well-referenced pages just how impactful Webb was during the Lavender Scare. This part is particularly important: 'That Webb played a leadership position in the Lavender Scare is undeniable. The only thing left up to historical debate in this matter is whether or not his heart was in it. Was Webb emotionally invested in the persecution of LGBTQIA+ people? Or was he simply trying to save face – by spreading homophobia and doubling down on the firings – following the public revelation of firings that had happened without probable cause? Either way, one thing is clear: he still did those things.'

Several emails in the FOI request reveal just how little many at NASA were interested in actually engaging with the community. In the end, NASA has given itself a report with which they can claim Webb's innocence, but, in reality, Odom's report doesn't exculpate Webb: Webb may not have personally fired a gay employee, but nor did he build the rockets, perform the calculations, or fly the *Apollo* missions to the Moon. So we cannot praise his leadership role during the lunar programme while absolving him of the responsibilities of the horrific practices going on in the agency at the time. Many research institutes, learned societies and universities, as well as many media outlets, have thus decided simply to use the acronym without honouring the man behind the name. A $10 billion monument shouldn't just be handed out without scrutiny.

At the same time, we cannot talk about cutting-edge astronomy without JWST. I am a firm believer that astronomy is for all. Our ancestors were already doing astronomy before we even had a term for it; it is a common heritage of humanity and should be enjoyed by everyone, regardless of nationality, race, sexual orientation, gender, or status. That requires being aware that, at the end of the day, science is made by people and it has biases and flaws, and those can be only addressed by being open and understanding of others' ideas. After all, the whole point of science is the flexibility of changing established ideas when improved ones are presented.

Research has shown that among professionals in STEM with the same education level, work effort, or job commitment, LGBTQIA+ people were more likely to experience professional devaluation, exclusion and harassment than cisgender heterosexual colleagues.[15] They often stay in the closet or are even *told to stay in the closet* if they want to progress in their career.[16] No one is asking straight astronomers to never mention their spouses, their kids, or anything of their private lives. LGBTQIA+ astronomers are asked to remain invisible. Bigotry everywhere is insidious and damaging as a whole, but in science it can also get to the point of being so ridiculous that it's almost funny. About one in three physicists are told that they shouldn't act 'too gay' by their seniors. Physicists pride themselves on their rationality and how they see the Universe and its properties as measurable and quantifiable, so if senior physicists believe that there is a threshold for when one is too gay to be a physicist, there must be a scale with which to measure it. How about from *zero* to *Elton John* in the bird costume on *The Muppet Show*? At what point is one 'too gay' to study physics? At 0.5 Elton? 2.76 Eltons? It's laughable. In a Universe full of amazing wonders, the obsessive policing of people's sexuality and gender is both pitiful and absurd. Just a few years ago, for the first time ever, we saw the edges of a supermassive black hole, and yet some people care more about how others are living their lives. It is their choice!

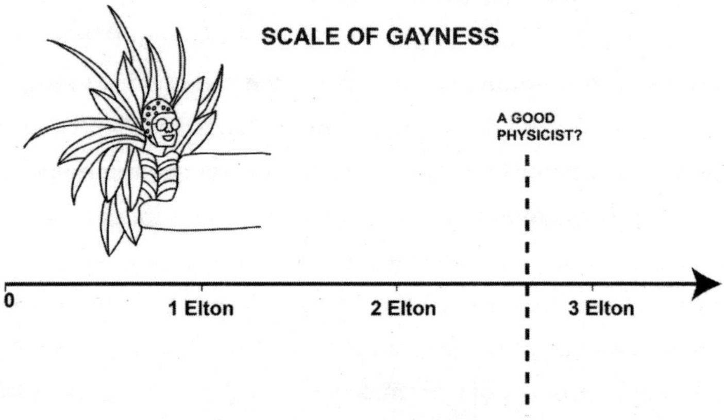

My fictional scale of gayness – apologies to Elton John for the drawing!

Alfredo & Chris Carpineti

Red Riding Hole

In the first chapter we discussed the revolutionary feat of taking that first image of a supermassive black hole, but supermassive black holes are of interest to infrared astronomers, too, as long as they are playing a game of cosmic peekaboo. These are the heaviest single objects in the Universe and yet they can easily become shrouded. They surround themselves with material, by nibbling or gorging on it, and when they gorge they are so messy that they become extremely bright. Still, depending on quantity, the material can shroud them from view. Well, that is if you are only looking in visible light. Dr Marie Wingyee Lau (they/them) will show us how we can see through the veil of dust and gas that hides them.

'A quasar is an extremely luminous phenomenon which describes the very active state of a supermassive black hole, and how the gas surrounding the supermassive black hole falls toward it,' explained Dr Lau as we talked over Zoom. 'This infalling gas heats up because of friction and releases energy in the form of electromagnetic radiation. This radiative energy released by a quasar is enormous and it blows away the gas surrounding it.'

Quasars are the most luminous objects in the Universe and the blowing away of the gas can create feedback that affects the whole galaxy. Enormous jets and galaxy-wide winds can spread across the galaxy, affecting events like star formation, and possibly even novae.[17] They are fascinating and well-studied and yet there is so much about them that we do not understand. Dr Lau is a non-binary immigrant from Hong Kong living in California, and we talked online in our respective early hours, the morning for me and night for them. The enthusiasm for solving astronomical mysteries was not dampened by the awkwardness of time zones. Dr Lau is a postdoctoral scholar at the University of California Riverside and stresses just how fantastic quasars are. There is a region around a supermassive black hole that becomes a powerful source of light so powerful that in some cases it changes its whole galaxy. And yet, even these extremely bright quasars can be invisible to us and end up shrouded in gas and dust. Phenomenal cosmic power, itty-bitty (and very dusty) living space. Lau's work focuses on those quasars shrouded in dust and how these incredible objects shape the evolution of the galaxies in which they reside. But studying them is only half the battle – first you need to find the quasars. The way to do it is by looking for oxygen using infrared light.

The quasars are not breathing like us, but they do interact with the oxygen in the gas and dust that shrouds them. All the energy that they emit ends up knocking off electrons from oxygen atoms. This is the same ionisation phenomenon we have seen in previous chapters, taking place during the first billion years of the Universe, with hydrogen becoming transparent under the effect of starlight. The ionisation of oxygen (when oxygen loses electrons) occurs when electrons from the solar wind slam into the atmosphere and create northern lights. In fact, the green colour of the northern lights is due to ionised oxygen. But this light that we can see with our naked eye here on Earth is redshifted in those distant galaxies, so you need infrared light to study them. Dr Lau uses infrared telescopes in their quest to quiz quasars about their role in galaxy evolution.

'Over time, a lot of massive galaxies stop forming stars, and there needs to be some kind of mechanism to crunch the star formation. This is usually proposed to be quasar feedback, but it has never been definitely observed,' Dr Lau explained. This is the focus of their research, a fantastic quest for the literally ultimate power in the Universe. 'A lot of astronomers think that quasar feedback is at the most powerful when they are in the dust-reddened state. So I'm trying to find the most powerful quasars driving the most powerful feedback. And by studying that, we can learn how galaxies, massive galaxies, stop forming stars.'

I can't help being in awe at this kind of research. We are specks of carbon on a speck of silicon orbiting an average star and yet here we are, not only being able to find the most luminous objects in the Universe, but finding them even when Mother Nature herself has shrouded them in dust. Our senses and their limits define us for sure, but our ability to constantly exceed them is truly a testament to the great ability of our species. In astronomy, that ability is at the service of finding out the crucial questions that vex us about the cosmos. How did it all begin? How will it end? And, of course, are we alone in the Universe? Understanding the origin of life requires working out what makes Earth suitable, if not special. For that reason, we need to find other worlds like it.

Looking for an Earth Twin

Infrared astronomy is massively helping with the quest to find other inhabitable planets. Statistically speaking, we can't be the only planet with life in the galaxy. There are just too many planets out there, hundreds of billions in the Milky Way alone. For sure, most of them won't have the right conditions for life, but thinking that life is an Earth-exclusive is pretty reductive. We will find life out there, although most likely what we will find is something much simpler than us, microbes of some sort. Maybe it is just a little exploration away. Microbial life could exist underground on Mars, or in the oceans under the icy exteriors of the moons of the gas giant planets. And, of course, beyond the solar system, in worlds that we might never reach. Studying those worlds in themselves gives us new insights into how the solar system came to be. To better understand Earth, researchers look for a twin of Earth, although the closest worlds found so far are more Earth-sized rather than Earth-like. Astronomers can find these faraway Earths by looking at their distance from their star. This tells us if they are in the habitable zone where water might exist in all three common states: gas, liquid and solid. But what we really need to make an Earth is an atmosphere. Once an atmosphere is detected, then researchers can hunt for biosignatures, the chemical signal of interesting gases in the atmosphere that could imply there is life there. Confirming the presence of these molecules is done in a specific way. The spectrum of light from a star is taken as an exoplanet passes in front of it. The planet's atmosphere filters its light. The light is absorbed by molecules in the atmosphere at specific infrared wavelengths, creating distinct patterns – carbon dioxide will have a specific signature, for example. Telescopes like Hubble have been using this approach to find interesting molecules in distant exoplanets. And now JWST has pushed that approach further, thanks to its infrared capability. Dr Jake Taylor (he/him) explained this game-changing ability of JWST. Jake is a very good friend and it is always a pleasure hearing about his work. We are both very much into exoplanets, and his excitement about JWST data is palpable. He is doing a postdoc at the University of Oxford, focusing on the atmosphere of exoplanets, so I couldn't ask for a better expert.

'There's just this inventory of molecules that we can look at that exist in the infrared. And so one of the benefits of having a space telescope is that it's really tricky from the ground to look at these atmospheres in infrared,' says Dr Taylor, highlighting how JWST has allowed astronomers to finally expand the molecular analysis of exoplanets' atmospheres.

As he explains it, the space telescope is delivering on the promise of its designers. Even in its first few months of operation, it discovered molecules that had not been seen before, such as carbon dioxide and sulphur dioxide on an exoplanet 700 light-years away.[18] It is extraordinary that we can tell that there is carbon dioxide in a world so far away. And the presence of that particular molecule is doubly exciting because carbon dioxide is considered a biosignature, so a *potential* sign of life.

The keyword here is 'potential'. Carbon dioxide doesn't guarantee life, and the exoplanet in this study, known as WASP-39b, is a perfect example of that. No life would want to be on a hellish world burning at 900 degrees Celsius. While the carbon dioxide news garnered a lot of headlines, it is the sulphur dioxide that tells us more about what's happening on that distant planet, including the fact that starlight is helping the creation of new molecules, a process known as photochemistry.

Dr Taylor continued:

> Sulphur dioxide is a molecule that exists due to photochemistry. The light from the star is hitting the upper atmosphere of the planet and it's causing a bunch of different chemical reactions that don't happen in other ways. We've never observed it before in an exoplanet and this is because we've never had the wavelength coverage to observe those molecular features that

> happen as byproducts or photochemistry. We're now seeing this process happen. We've confirmed it. And this tells us that other planets outside the solar system have photochemistry. This is a very key process because on Earth's atmosphere, this is how we obtain ozone. And ozone is one of the parts of the atmosphere that keeps us protected from ultraviolet light.

Photochemistry and the presence of carbon molecules in the atmosphere provide evidence that our planet and the solar system are not special flukes. This is what happens to planets everywhere in the Milky Way. While WASP-39b might not be the most hospitable place, other worlds might be very different. A world like Earth could be just a lucky observation away, but we are still relying on a lot of luck to find it. There is a lot of space to cover and only a small number of telescopes working on a tight schedule. This creates a bias towards systems with planets that orbit much faster than Earth. If you are waiting for a planet to pass in front of its star, it's much easier if they do that every few days than every eleven years. Finding a solar system like our own would take centuries, since Neptune goes around the Sun every 165 years and you would want a few detections to be sure you really see a planet blocking the light of a star; you can imagine the waiting time.

You Can't Make a Planet Without Breaking a Few Stars

As I mentioned at the beginning, we use light beyond the visible to discover what we can't see, or to learn more about the objects and events that we can see. It is hardly surprising that we encounter stars, galaxies and black holes over and over again, but there are only a few observatories and their time is limited, with astronomers making competing proposals about what is worth pursuing. Some fields of study were available decades ago, but nobody had the winning argument to get access to the right telescopes that could deliver the breakthroughs.

One example that surprised me was that the systematic observations of supernovae with infrared light is a new field, younger than most social media – less than ten years old. Dr Kaew Samaporn Tinyanont (he/him), from the National Astronomical Research Institute of Thailand in Chiang Mai told me more about it. The advantage of using infrared, as we have seen, is the ability to see through dusty areas – so the infrared light of these supernovae is less affected, delivering more insights. But there is also the emission of dust and molecules that form as a supernova cools. As those molecules and grains shine in infrared, they tell us about where we might have come from, and they could solve the Hubble tension we encountered earlier in the book. When we have a bright stellar explosion, the explosion ejects a lot of matter, some of which slams into the material that surrounds the exploding star. We can keep track of these events to see how they change over time: it tells us what came before and so much more.

'Usually, several months after the explosion, the ejecta has cooled down enough that molecules start to form. It starts with carbon monoxide, which is like a very simple molecule. So these things form first and later on you have stuff like silicon oxide and some other more complicated molecules,' Dr Tinyanont explained.

I had him on my list of interviewees for a very long time, but when I first contacted him, he was in the process of moving from

California to Thailand and he did not have time. It was over a year later, when I delivered a hybrid talk in Thailand, where we reconnected. Luckily, I didn't let him get away a second time. He also delivered some important titbits about the role of supernovae in the history of the Universe. One crucial point is this dust we keep talking about. The dust-shrouded quasars, or reddening galaxy, is all well and good. It's important. From that dust, you eventually form planets, as we have seen. But thanks to JWST and ALMA, we have started to see that there might be a bit too much dust in some of the earliest galaxies.

'There have been some observations in the far infrared to submillimetre showing that these really, really distant quasars and galaxies have a lot of dust in them. And these are at high redshift when the Universe was only like a few hundred million years old,' Dr Tinyanont continued.

So where is the dust coming from? In the local Universe, we believe that the main source is massive (but not too massive) stars that have evolved beyond the main sequence – so they stopped burning hydrogen like the Sun. This might not be the only source, but it is considered a major player. However, in the early Universe, they have not had time to evolve ... Researchers believe that it is supernovae creating this dust, and they are using infrared telescopes for these events to better estimate just how much dust they produce. Currently, there are no certain answers.

There is also another morsel of information in our conversation – understanding supernovae across the spectrum of light makes the models better. These models are crucial to estimating the properties and luminosity of supernovae, including those used to estimate the distance of galaxies.

This brings us back to the exciting and controversial topic

of the Hubble Tension. In the microwave chapter we saw that astronomers are struggling to estimate how fast the Universe is expanding. Using the cosmic microwave background they get one number, while measuring the distance and redshift of galaxies with Hubble they get a different one. JWST observations are now being employed as well, with even more controversy. Several approaches still reaffirmed the tension, but in mid-2024 work by Wendy Freedman and colleagues claimed[19] that maybe there is no tension at all. Freedman did not use supernovae but, rather, three different types of star. The different observations stated that the tension is alive and well.[20] There has been some back and forth between groups about what this or that dataset from JWST tells us. It is the cosmological question of our times. The answer continues to elude us, but it's not JWST's fault. The telescope is going to be crucial in providing clarity, since its high-resolution vision can penetrate some of the dust of galaxies and see not only the stars but also highlight different type of dust. And so, in infrared, we seek dust and we avoid it. It's a veil, a resource and a building block.

Quantity Without Forgetting Quality

It's no surprise that big numbers are referred to as being astronomical – astronomers are all about the large numbers. Soon we will have many more infrared targets that will not only help cosmology and understanding the Universe but also how we find planets. In microwave and submillimetre, we have seen that planets can be observed directly if they are far enough away from their star (or if their star is dim enough). Alternatively, by looking at the star, it is possible to measure the small gravitational pull of the planet on its star, since that creates a little wobble (changing the starlight). But there is also an approach called gravitational microlensing that is worth exploring; much like the gravitational lensing approach from the beginning of the chapter, this, too, uses gravity as a magnifying tool. Yet the effect of a planet on a star is a lot more modest – so it is micro, but it is very relevant to infrared astronomy because it will be employed by one of the next big infrared observatories: the Roman telescope.

This telescope is named after Dr Nancy Grace Roman, the first female executive at NASA and the agency's first chief of astronomy, honoured for her founding role in planning the Hubble Space Telescope. It has an impressive camera, 100 times larger than the imaging camera on Hubble, which will allow astronomers to map larger areas of the sky very quickly, or smaller areas (but still much larger than Hubble and JWST) more often, delivering more insights into varying phenomena, from the expansion of the Universe to planets. Astronomers plan to use this incredible tool to conduct a vast census of hundreds of millions of stars in the core of the Milky Way, and the expectation is that they will be spotting exoplanets, thanks to microlensing.

In this microlensing approach, the foreground lens is the interesting object. A gravitational lens will make a background object appear brighter. The smaller the mass of the foreground lens, the more modest the effect, but with a good enough telescope, like Roman, it is still measurable. Let's imagine a distant star (A) and then imagine a slightly closer star passing between us and A. This new star (B) has a planet orbiting it. As it moves through the line of sight, the brightness of A increases due to the warped space-time created by B. B's planet also warps space-time, by a much smaller amount, but it also creates an increase in the brightness of A, And that's how you'd detect this planet. But as we say in Italian: tra il dire e il fare c'è di mezzo il mare. Between theory and practice, there is an ocean.

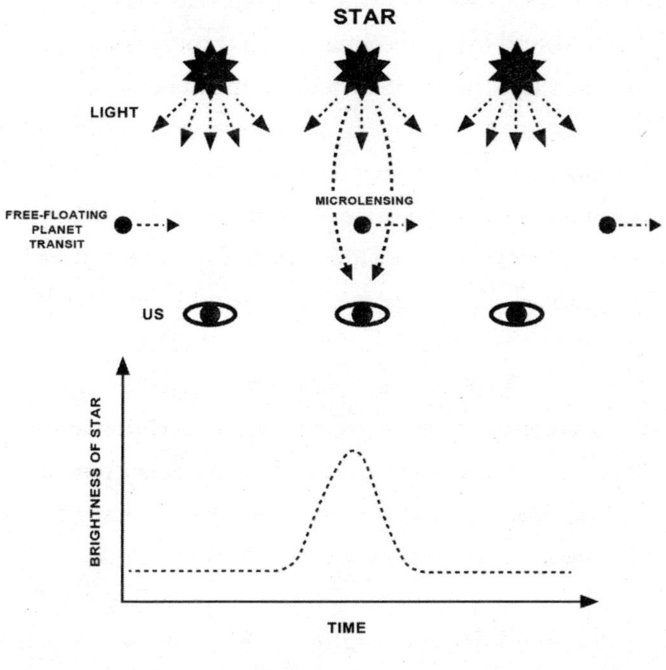

Schematic view of how a free-floating planet can create a microlensing event.

Alfredo & Chris Carpineti

'Almost all the methods we have so far that can do this [detect exoplanets] in particular, Kepler, which detected thousands of planets, are only sensitive to planets that are relatively close to their host stars. So we don't know what planetary systems look like in regions where Jupiter, Saturn, Uranus, and Neptune live,' explains Professor Scott Gaudi (he/him), the Thomas Jefferson Chair for Space Exploration at Ohio State University, Columbus. Gaudi is one of the principal investigators in the Roman telescope, and I felt very lucky to get the chance to talk to him. Gaudi is not only an award-winning scientist and co-discoverer of the hottest known exoplanet (KELT-9b), but he has worked tirelessly to support marginalised minorities in astronomy and challenge harassment found in our discipline. In our conversation, Gaudi explained just how revolutionary Roman will be:

> The problem with gravitational microlensing is that is a very rare phenomenon. So you have to look at a lot of stars, and the location in the Universe near us that contains a lot of stars is the centre of our galaxy: the galactic bulge. But there's a lot of dust along our line of sight to the galactic bulge which obscures the optical light, but not so much the infrared light. So by going to the infrared, we can see more stars, and therefore detect more microlensing events, and therefore detect more planets. And the other aspect of it is that it allows us to study lower mass stars which put out more of their light in the infrared.

This method will deliver insight into populations of planets we currently know very little about, including planets without stars – the so-called free-floating planets or rogue planets, a class of planets that are believed to have been kicked out from their original planetary system due to gravitational interactions with other planets or passing stars. It is believed that free-floating planets are very common, but they are very cold, so we can't see them directly using visible light.

The Roman telescope will also gather further data on more quasars, finding objects that so far have been too dim to be spotted. Together with Euclid,[21] a mission by the European Space Agency launched in 2023, they will take measurements to determine how the Universe has expanded, and try to understand the nature of dark matter and dark energy. And there is always more. The 2030s will see the launch of the Origins Space Telescope. This will focus once again on far-infrared wavelength, looking at further, cooler objects in the Universe. It is a survey telescope, so it will conduct many different types of observations, from distant galaxies and galaxy evolution to the Milky Way and the material between stars – the so-called interstellar medium – from the formation of other planetary systems to studying our own. It will attain sensitivities more than 1,000 times greater than any previous infrared telescope.[22] Dr Lau has great expectations of both Origins and Roman:

> If you go down in luminosity, you study things that are more representative and more informative of the general galaxy evolution picture. And that's what Roman can help us achieve. So currently we do have some all-sky or wide-area surveys done from the ground. But Roman is going to go much deeper. So we can find fainter red quasars and then with Origins, you push even fainter.

Infrared astronomy encapsulates so much of the opportunity that seeing the Universe with other wavelengths allows. There are things that we know full well exist, but we would not be able to tell without this portion of the light spectrum: planets and their atmospheres, stars, brown dwarfs, supermassive black holes, and even whole galaxies. They are there living their best lives, but we would be completely ignorant about them and their characteristics if we hadn't taken a step beyond what we could see and worked out what was causing this heat that we feel. As Dr Lau explained, the future of infrared astronomy is about pushing the limits of what we can see of the very faint, so we can't literally say that the discipline's future is bright, and it is, in my opinion, incredibly fascinating.

In our journey beyond the visible, we have crossed the wavelengths which are longer than the visible light we see. This light has shown us that there are things invisible to us in the Universe, some too faint and distant, others bright but shrouded. Radio waves, microwaves and infrared in particular have opened the door to finding the answers to the biggest questions, from the origin of galaxies to the possibility of life elsewhere in the cosmos. We are halfway through our journey, and you are coming out of the cave to squint at the sunlight. The light of the objects we are about to meet can be seen but there is more hidden around it. We will start with ultraviolet light at the opposite side of visible light on the spectrum. Just because something is bright and visible, it doesn't mean we are getting the whole story.

— CHAPTER 4 —

ULTRAVIOLET

The Blinding Limelight

For like a shaft, clear and cold,
the thought pierced him that in the end
the Shadow was only a small and passing thing:
there was light and high beauty
for ever beyond its reach.

—J. R. R. Tolkien

Mattina
M'illumino di immenso.
[Morning I'm illuminated with immensity]

—Giuseppe Ungaretti

Thanks to radio waves, microwaves and infrared light we have discovered that there is a whole Universe of invisible, shrouded and cool objects. Cool both in temperature and also because they are awesome. But some like it hot, and it is fair that we also answer the burning questions you might have about all the cosmic inhabitants that sizzle and dazzle with their light. That path will take us into the ultraviolet portion of the electromagnetic spectrum. The stars of this chapter are, well, stars. Directly or indirectly, they are the source of ultraviolet light, something not to be taken lightly. Their impact, as we shall see, is cosmic. You've probably already experienced ultraviolet light as an enemy of sorts – it can burn us, give us skin cancer, damage our DNA, and it does make our skin wrinkle and age. I know, I know, this is not a great introduction for ultraviolet light – we've all got beef with it – but let me tell you how important it is not only for astronomy but also to protect our way of life, despite the known dangers. We are about to meet objects and processes (and people) that are bright. Some extremely bright. Often we can see them perfectly, but this apparent visibility is limited and limiting. The light we see can sometimes be so bright that the most interesting details are missing. You have visibility but only certain aspects are shown.

Ultraviolet light is important for two reasons, one small and one big, but both are connected. The small reason is the hydrogen atom. In its simplest form, you can imagine this atom being made of a proton - a positively charged particle - orbited by an electron, which is negatively charged. The electron orbits the proton in a specific region. If you give it a lot of energy, you can rip the electron, as we saw in the ionisation and reionisation discussion in the radio waves chapter. If you give the electron some energy and make it jump to a higher-energy region, it will eventually jump back to its ground state, releasing a photon. This is the principle behind the lasers and the masers we saw in the microwaves chapter. The jumping to the ground state happens by emitting one photon of a very specific wavelength, depending on the excited region from which it's coming. And, for hydrogen atoms, they are all in the UV portion of the electromagnetic spectrum.

Hydrogen is the most common element in the Universe, so its ability to emit UV is already enough to make us interested in the subject. But it is even more interesting for us that hydrogen needs to absorb ultraviolet light to lose its electron and become ionised. That's what a lot of hydrogen in the Universe is like: ionised. Let's quickly revisit what happened during the epoch of reionisation, what happened in the first 1 billion years of the Universe. Following the emission of the cosmic microwave background, you just had regular hydrogen. Regular hydrogen is opaque, a misty gas stretching across the cosmos. Ionised hydrogen is transparent, so ultraviolet light was responsible for making the Universe transparent. The source of the ultraviolet light? Massive stars (and possibly black holes); stars, especially those that are massive and hot, emit a lot of UV. I spoke to Professor Sally Oey (she/her) whose area of expertise is massive stars and their effect on galaxies.

'Massive stars emit most of their light in the UV. So that means that they are responsible for a huge fraction of the ultraviolet light that we see in the Universe,' Professor Oey stated matter-of-factly. We will learn more about them and about Professor Oey shortly. This information was necessary to guide us on our journey into the UV, but we need to start closer to home, in our little corner of the Universe and our little star: the Sun. After all, it is studying the Sun that pushed the development of ultraviolet astronomy.

The View Beyond the Stratosphere

The development of UV astronomy is tied to the space race. Because, *unfortunately*, we can't do ultraviolet astronomy from the ground. That emphatic 'unfortunately' is your little window into the minds of astronomers. Sometimes we complain about features of our planet that are in the category 'things crucial to life on Earth'. Just as air and water vapour make infrared astronomy difficult on the ground, so the ozone layers that protect all the natural world make ultraviolet observations from down here a non-starter. We need to go to space, and the first space telescope looking into ultraviolet light was in the hands of an astronaut. Literally. There was a human-helmed orbiting solar observatory that delivered important observations of the Sun in both ultraviolet and X-rays, kickstarting both branches of astronomy. The pivotal astronaut behind that instrument was Ed Gibson.[1] Since then, astronauts and agencies have delivered incredible instruments to space near and far from our planet, bringing eyes that can see ultraviolet light from the cosmos. Just like Gibson before us, our focus will be on the biggest source of ultraviolet in our vicinity: the Sun. Most of the ultraviolet light from the Sun doesn't come from the visible surface of the Sun – aka the photosphere – but from a thin layer of invisible plasma that lies above it – the chromosphere. It is the main source, although the further outer layer of the solar corona can contribute as well. Here we need to introduce one of the most complex and strange mysteries about the Sun: the chromosphere and the solar corona are hotter than the surface of the Sun. To quote *Knives Out*'s Benoit Blanc, *it makes no damn sense!* But it compels thousands of scientists though. The solar corona is in the millions of degrees. Clearly, it doesn't make sense. How can something further away from the source of heat be hotter? And yet it is.

Looking at the Sun in ultraviolet and X-rays provides insights into these extremely hot and active regions, where researchers are getting close to finding answers to this enduring mystery. I was lucky enough to have an insider's view of this quest from someone working on one of these missions: Dr David Williams (he/him) of the European Space Agency.

'There's so much new information in looking at the ultraviolet light of the Sun because it's not competing with anything from the surface. It can shine by itself, it's loud, it stands out,' Dr Williams told me. Dr Williams is a dear friend but, more importantly for this discussion, he is one of the scientists in the Science Operations Centre for the European Space Agency's Solar Orbiter mission.

Solar Orbiter is studying the Sun as never before with incredible observations of its properties. Just last year, it provided the first ever images of the polar regions of the Sun, which had never been seen before. Dr Williams has a multi-class role, from scientist to technician, engineer to spacecraft secretary, to helping coordinate and make observations happen in a way that is successful and safe. Talking to him about the importance of looking at the Sun in UV is both a pleasure, and, excuse the pun, illuminating. He begins to compare the way we look at the Sun to a car radio and how you have to turn the dial to find the right station. While I was nodding along to this analogy, we both stopped and looked at each other, realising that car radios have not come with a hand dial in decades. A more modern analogy might be choosing the right filter for a photo, which would make certain details pop. What we are trying to say is that by picking the right wavelength in the ultraviolet range, you can see below and above the surface, usually in places where something exciting happens.

'You pick up all sorts of different information, depending on exactly which wavelength you choose. Some wavelengths of light will show you where the magnetic field is concentrated at the surface of the Sun. And it's concentrated in patches of all sizes, like bigger than planets, down to a few city sizes across,' Dr Williams explained, talking about sunspots, the darker and cooler region of the Sun's surface.

> There are ultraviolet wavelengths that will let you peer deeper below the surface because that's where the magnetic fields are concentrated. There are wavelengths that will let you look at the vast magnetic loops that cover the surface of the Sun. There are wavelengths that will let you see jumping pillars of gas, which we call spicules, which are constantly being sort of bounced around on the surface of the Sun. And really, depending on where you look, you get very different information. It's an extremely powerful toolkit.

Solar Orbiter and so many other missions would not exist without ultraviolet observation. They would not make sense without it. Their goal is to allow us to understand the Sun. We have a rough idea of what goes on, but so much, like the corona heating mystery, is indeed still a mystery. And if we can't understand our own star, which is 'around the corner' celestially speaking, how can we hope to understand any other star in the Universe, especially in our search for stars orbited by habitable planets? Ultraviolet observations keep an eye on the Sun not just for pure science but also for planetary protection. The Sun is constantly releasing particles into interplanetary space – the solar wind – but those regions of activity can also release bright flashes of light and showers of high-energy plasma that pose a risk to communication and technological infrastructure on Earth. This is the space weather that connects Earth to the Sun, and this connection and its changes can be seen, appreciated and monitored by ultraviolet observations. We can't stop the Sun from spewing plasma into space, but we can prepare ourselves for potentially dangerous events that could disrupt all our tech. But it is not all bad news; that same flow of particles gives us the aurorae borealis and australis – the northern and southern lights, respectively.

The Curtains Shaking Over Many Worlds

The aurorae can be seen with the naked eye at higher latitudes (and at lower ones if the Sun is in a particularly active portion of its eleven-year cycle) and, as far as my personal preferences go, they are one of the most incredible spectacles I have ever been lucky enough to witness. And while the earthly ones are in visible light, it turns out that the Universe provides a wide selection of aurorae. Astronomers have observed ultraviolet aurorae across the skies of almost all planets in the solar system (and even things that are not planets at all). Aurorae usually work in the following way. Charged-up particles from the Sun interact with a planet's magnetic field, following its lines that go from one pole to the other. These particles, usually electrons, eventually slam into the atmosphere, and, depending on the energy of the charged particles and their interaction with the magnetic field, they can penetrate more, slamming different gases – nitrogen or oxygen on Earth – and produce different colours. They do not remain an exclusive of visible light. I sat down virtually with Dr Affelia Wibisono (they/them) whose PhD focused on the

X-ray emission from planets. We will, of course hear more from them in the next chapter, but their expertise is needed here as well, as ultraviolet aurorae and X-ray aurorae happen over the same worlds.

'Detecting aurora on a planet tells us that that planet has an atmosphere, and likely a magnetic field. Auroral emissions are a byproduct of charged particles interacting with the gases in a planet's atmosphere. For the Earth, Jupiter, Saturn, Uranus and Neptune, the charged particles are accelerated and travel along the planets' magnetic field lines and enter the atmosphere around the poles, hence why their aurorae are seen in the polar regions,' Dr Wibisono told me. They are a public engagement officer at the Institute of Physics.

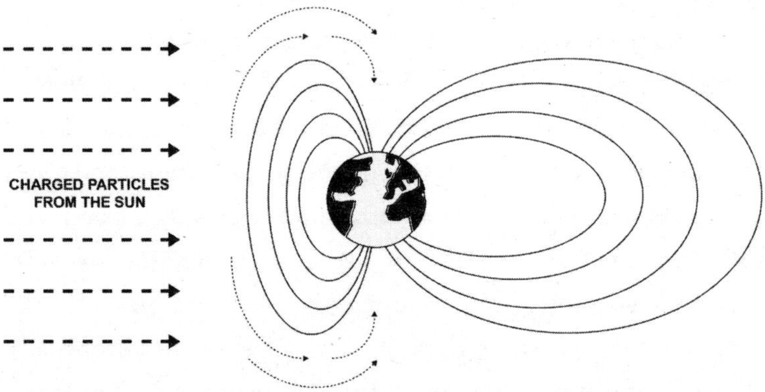

A simple schematic showing how aurorae form. Electrically charged particles from the Sun are pulled towards the poles by the magnetic field of the Earth and slam into the molecules in the atmosphere.

Alfredo & Chris Carpineti

The *Voyager 2* probe recorded ultraviolet aurorae on Uranus and Neptune, but that was a flying visit, as the spacecraft just went past the planets without entering into orbit. Hubble was able to only see Uranus' aurorae in follow-up observations.[2] Neptune's ones were only recently seen by JWST,[3] so technically infrared aurorae… but more are expected! Hubble has also cast its UV-sensitive eye to Saturn,[4] spotting its ultraviolet aurora several times. The aurorae on Earth and Saturn are very similar, just having a different level of energy and atmospheric composition. Jupiter is a whole other can of celestial worms…

'The giant planets have bright ultraviolet aurorae because their atmospheres have a lot of hydrogen. The precipitating electrons excite atomic or molecular hydrogen in the atmosphere,' Dr Wibisono explained before we got on to the peculiarities of Jupiter. Aurorae are usually dependent on the effect of the solar winds, so they wax and wane as the Sun increases or decreases its activity. But not for Jupiter. They are permanent due to the planet's extreme magnetic field which is 20,000 times stronger than Earth's own, and its moons, which affect the plasma surrounding the planet, as well as releasing more particles in such an environment. So, Jupiter has ultraviolet aurorae that not only follow the planet's own magnetic field lines but are also caused by its moons, Io and to a lesser extent Europa, Ganymede and Callisto, which slow down plasma, creating spots of aurorae on Jupiter.[5] And the aurorae go both ways, with aurorae (even visible ones) having been seen around these moons. Currents are actually present, connecting the volcanic moon Io and Jupiter. The solar wind can also contribute to delivering occasional bright spots in the big oval of the main constant aurora.

Thanks to ultraviolet light, we can track and better understand the complex magnetic and electric behaviours around Jupiter. It is a place of intense radiation, which may impact the habitability of Europa, an icy moon with an active liquid ocean under its thick ice shell, as well as Ganymede and Callisto, which might also hide deep oceans. Missions to study these moons and their potential for hosting life will reach them in the early 2030s and will also be equipped with UV detectors. Given ultraviolet light's ability to kill simple life forms and damage DNA, it is important to keep an eye on it. It's not just about our search for life on the icy worlds orbiting the gas giant in the solar system – Jupiter shows that Earth's aurorae are not a one-size-fits-all model, and there are other planets and minor worlds to consider.

As for Mars, we have a large fleet of orbiters and robots, so we get some pretty interesting details about this planet and its behaviour. The Red Planet hasn't got a magnetic field any more, but it used to have one as strong as Earth's. Evidence of this magnetic field is literally trapped in rocks, and these magnetised regions can influence the formation of aurorae,[6] because even without a global magnetic field it still gets shiny ultraviolet aurorae – but they can be discrete, localised on specific regions rather than around the poles, and they can be patchy, depending on the ability of the solar-wind protons (it's not just electrons escaping the Sun) to penetrate the Martian atmosphere. Patchy proton aurorae have been detected across Mars, providing some answers to their formation, as well as more questions. Magnetism is, after all, complicated.

Venus, too, without a permanent magnetic field, experiences aurorae. Its night side experiences ultraviolet aurorae, possibly due to a rain of electrons, and these aurorae are influenced by solar flares and/or interplanetary shocks. The smallest object to have been observed having ultraviolet aurorae so far is comet 67P/Churyumov–Gerasimenko.[7] The gas that surrounds the comet's nucleus, the coma, is excited by electrons in the solar wind. The electrons break apart the water molecules, releasing hydrogen and oxygen and making the gas glow. These were confirmed by the European Space Agency's *Rosetta* mission. Space missions have ultraviolet instruments because they can be used to analyse gases and ices. They can also reveal the presence of these beautiful displays of glowing gas. For comet 67P, the glow of its gas had previously been linked to photons interacting with the gas. The observations from *Rosetta* provide the details of its true nature. So far in the solar system, we have seen UV light being used to see additional intriguing titbits that are often lost to the natural visibility of the objects in question. Details that at first, or historically, might not have looked fundamental but are now appreciated as being crucial parts of the rich tapestry making those celestial bodies what they are.

'Studying the Universe in different types of light lets us understand the Universe more completely because different processes and different objects emit light from different parts of the electromagnetic spectrum,' Dr Wibisono continued. 'Ultraviolet light gives us information about the hottest, largest and youngest stars. So ultraviolet observations of the sky tell us which galaxies have stars forming within them and also where in those galaxies these baby stars are.'

Twinkle Twinkle Extremely Hot Massive Globe of Plasma

To the stars is where we go next. Professor Sally Oey from the University of Michigan is the perfect expert for this journey. In our conversation, she laid bare for me everything we know of massive stars and how they affect their local environments and galaxies as a whole. Maybe a single star won't change its galaxy, but the cumulative effect of their light will be felt eventually.

The one that we are most familiar with is, of course, the Sun – we know it well but not completely, as we have just seen. In fact, given that we have an incomplete understanding of the whats and hows of the source of all life on Earth, our understanding of distant stars is even more patchy. But limited understanding is a huge leap from ignorance, and every day we are filling in the gaps.

The Sun is an average-sized yellow dwarf. There are smaller stars and much bigger ones. Anything with a mass eight times our Sun or more is a massive star. And they burn hot and bright. As Professor Oey teased at the beginning of the chapter, these stars emit most of their light in ultraviolet light and most of the ultraviolet light in the Universe comes from them. Their life and light are very important, as we shall see how they change the Universe. The value of 'massive' starting at eight times the mass of our Sun is not arbitrary. There are plenty of discretionary definitions in science, but we are not considering a group of stars as massive on a whim. They are in this category because of something they have in common: the manner of their death. The eight solar masses threshold tells us the fate of the star in question. That's how heavy a star needs to be in its regular life, the so-called main sequence, to then end up in a supernova. But every moment in the lives of these stars will be poignant, so let's start in chronological order with their birth.

Stars are born in vast reservoirs of hydrogen which are known as massive molecular clouds but more often than not we call nebulae. Nebulae from which stars form are known as stellar nurseries, and on average they are several light-years across. Many stars are born in such nurseries, but you need to have the right conditions. It's not a free-for-all. Counterintuitively, for stars to begin to burn, the gas from which they form needs to be very cold; the colder the better, actually. The reason for this is simple: you need the gas to condense. Colder gas has less pressure pushing it outwards, making it condense. Colder gas occupies a smaller space, and if your goal is to make a star, you need the gas to be losing energy, to become colder. That makes the gas cloud shrink. As the gas becomes denser, its mass gets too big for that region, and the relentless pull of gravity takes over, turning the cloud into a star. The temperature drop is seen as a crucial step in making stars, but it is not the only way to have this overdensity. This can also happen if the mass of the region is just too big, but you can't get the gas cold enough. Gravity will still take over. Another way for this to happen is if a disturbance creates an overdensity of gas in a specific region. Gravity is only attractive, so extra density will attract more mass and cause the collapse. These are all wonderful ways to make a dense spherical object, but we are not calling that a star just yet – a few more things need to happen to that collapsed object to make a star. With the collapse of the cloud, you obtain a protostar. This object is a fledgling that will continue to steal material from the gas cloud and grow bigger. This particular event isn't fully understood and remains a bit of a cosmic mystery… These protostars are embedded deeply in the stellar nurseries and we can't yet see through them. Thanks to infrared light and microwaves, we have started to see more details of this process, but there is also a lot of guesswork. Our best description is the following: protostars are different from stars in their heart. I am not being poetic, claiming that only the protostars with the purest of hearts can become stars. I am talking about nuclear fusion: the way that stars are powered. Due to the intense pressure and heat at the centre of a star, hydrogen atoms are combined and fused together in what becomes a helium atom. This process releases a lot of energy, some of which, over time, escapes the star and can travel all the way into our eyes. But protostars don't fuse hydrogen at this stage. They are just getting started. Despite the lack of fusion, they are quite active. As they bring in material from the stellar nursery, this accumulating gas hits their surface, producing shockwaves, and releasing energy that we can observe. When this period is over, the protostar will once again collapse under gravity, and as the object contracts, it will lead to a point where it gets the right conditions to fuse hydrogen at its core. And voilà, a star is born.

Not all stellar nurseries are the same, both in time and space. There have been periods when this process was much easier, faster and more successful. Nowadays, in our galaxy, a star the size of the Sun is born every year. But billions of years ago, the Milky Way was experiencing perhaps 1,000 times this rate of star birth. It was a swirling mess brimming with stellar nurseries. Massive stars form in these environments as well, but not as many as the smaller stars, of course. You get a few big ones and a lot of smaller ones. Though their numbers are small, their impact can be big. Their ultraviolet emissions affect the very gas in the stellar nurseries, making the gas and the whole nebula glow. When you are studying these stellar nurseries, you will often find these glowing regions. By studying these glowing regions you can gain a lot of information about the current state of the galaxy. How much gas is available to make new stars? What kind of stars are forming? Where are they forming? It is a very important way for astronomers to study star formation in galaxies in a wider sense and understand the evolution of that specific galaxy. This is done simply by seeing the effect of ultraviolet light on hydrogen gas, and by using instruments working with the deep red and infrared part of the spectrum, as you can see this effect clearly.

'Even though you're not directly detecting the ultraviolet light, you are still indirectly seeing the tracers of where these massive stars are through their nebular emission,' Professor Oey continued. 'If we are studying star-forming galaxies, that means that we can basically trace where star formation is happening, how much of it is happening, which galaxies are doing this star formation. And that's a very valuable way to be able to identify what's going on in galaxies.'

The effect of these massive stars doesn't end here. As stars evolve into the main sequence we mentioned above, their light will do more than just make the gas glow. It will break the gas apart, ripping electrons from hydrogen atoms (we have seen the phenomenon before, but here we call it photoionisation). The gas will be heated and pushed apart. We now know this heat can kill new stars and stops future ones from being born in the same nebula. Stars like our Sun will burn their hydrogen for billions of years, eventually running out of it in their core. They will experience a little collapse and contraction, which will increase the pressure and temperature at their centre, just enough to fuse helium into heavier elements. This change in fuel also makes a clear change in appearance. The star will swell up until it is no longer in the main sequence: it is now a red giant.

A red giant is very hot and enormous in size compared to the original star. The Sun might reach all the way to the Earth's orbit when it turns into such an object. This makes the outer layers of the star a lot less dense, and more loosely held together by gravity. The star will eventually run out of helium, too, and its core will collapse into a much denser object. Then its outer layer will blow away, revealing a white dwarf. Both helium-burning stars and white dwarfs emit a lot of ultraviolet light too. Ultraviolet light is not for the whippersnapper galaxies full of bright young things; it is also for the older galaxies, full of evolved stars.

For massive stars, the process of going from hydrogen to helium is much shorter; in some extreme cases, it's in the millions of years, but they can go on and fuse more than helium. Once they exhaust that fuel they will move on to carbon and oxygen. Some stars will go past that stage, and go on fusing even heavier elements. It is thanks to those stars, those evolutionary processes, and their eventual explosive death, that we are here today. They made the elements that make our planet and ourselves. Among the many evolved massive stars, we find the peculiar and heterogeneous class of Wolf–Rayet stars. They have distinctive ultraviolet light, an enhancement of elements heavier than hydrogen, and are some of the hottest stars known, with surface temperatures between four and forty-two times that of our Sun. One of the most massive known stars, R136a1,[8] is in this class. That's a star of almost 200 times the mass of the Sun. We are not talking about stellar peanuts here – we are talking about the hottest and most massive stars in the cosmos. And yet if you were to look at any of them with visible-light telescopes, you wouldn't see anything special. Their light is mostly in the ultraviolet part of the spectrum. That's where things are interesting: one can easily trace them in UV and work out all their incredible properties.

There is a very important reason to study Wolf-Rayet stars. They can't be 'born' like that, meaning that they evolved into that classification thanks to mass loss from prodigious stellar winds. But, intriguingly, they can be made if massive stars are in binary pairs: the companion, or secondary, can strip the primary star of its outer material, creating a Wolf-Rayet star, but the stolen material can heat up the secondary, also turning it into a Wolf-Rayet star. Most massive stars are believed to be in pairs in the Universe; by studying these special binaries it is possible to trace where binary stars are, what their properties and their numbers are, and so on. Thanks to the ultraviolet signatures, astronomers can track these stars and answer these questions. These observations provide crucial insights into the evolution of massive stars before they die, and give the Universe the material to make life and new stars and how they change the galaxy for the stars that form planets where life can evolve. By looking at the Universe in ultraviolet, or at least considering the ultraviolet impact, we can find out where the stars of any mass are born and where many stars are dying or have died. For massive stars, we also see their evolution and how they are shaping the interstellar regions around them both in their birth, in their evolution, and in their explosive deaths. That is not all.

As Professor Oey explained:

> Massive stars are very energetic; they are responsible for these evolutionary processes in galaxies. They explode, and they

> produce a lot of ionising radiation. They're extremely luminous. They are responsible for synthesising the heavy elements in the Universe. Massive stars are responsible for feedback in galaxies, meaning that they blow away gas through stellar explosions, strong stellar winds, and so forth. This feedback therefore clears out the gas. And that means that the ultraviolet light itself from the massive stars is directly able to escape from galaxies. So that process is very important for lighting up the gas that surrounds the galaxy, the circumgalactic medium, and the intergalactic medium. Currently, we think that it parallels what happens in the early Universe.

Professor Oey's group is looking at smaller compact galaxies which are doing just that. The parallel with the early Universe is to the epoch of reionisation we encountered at the end of the radio waves chapter. Massive bright stars and supernovae, likely also the first quasars, contributed to the reionisation of the Universe. To study the process at the actual time, you need longer wavelengths due to the expansion of the Universe, as we have seen. But to understand the processes we need the details that we can only get when we study those galaxies, those black holes and those massive hot stars nearby.

A Source of Iron and a Milestone

Once again we encounter supernovae, and once again we are getting a different aspect of these incredible explosions brought to light with the light we can't see. In particular, we are going to look at a specific kind of supernovae known as Type Ia (pronounced one-a). These supernovae happen in binary systems where one of the two objects is a white dwarf. As we have seen, a white dwarf is the leftover product of the evolution of stars like the Sun. This is not the end of the road for all white dwarfs. If they do have a companion, they can steal material and eventually become massive enough to explode.

A different approach is for a white dwarf to collide with another one, becoming too massive for the quantum mechanical properties that keep white dwarfs from collapsing. That mass, known as the Chandrasekhar limit, means that these thieving white dwarfs have a fairly consistent peak luminosity. As we have seen, this is like seeing a distant lighthouse and knowing exactly the type of light bulb installed in it. Not only do you know something is there, but you know how far away it is. And it's probably thanks to these explosions that you are alive. The iron in your blood, carrying oxygens to your cells, comes in large part from them. To better understand the importance of Type Ia supernova, I put myself in the hands of an expert, Dr Yaswant Devarakonda (he/him), who worked on this specific subject during his academic career.

Dr Devarakonda explained to me during our chat:

> They are really interesting objects for basically two main reasons, depending on what you want to look at. Type Ia supernovae are the origin of a lot of elements in the Universe. For instance, at the moment we believe that most of the iron in the Universe comes from Type Ias. Understanding the origin of the elements and the origin of stuff in the Universe, that's one thing that people are really interested in. The other thing, the thing that

> I'm more interested in, is the cosmological properties of Type Ia supernovae: they are very excellent at being standard candles.

We have seen that standard candles are so important in cosmology, so you would expect these objects to be profoundly studied. This is not exactly the case. Despite the lack of many observatories focusing on the ultraviolet, the data they get is very important for studying the creation of heavy elements, such as all the metals produced in these events. We need ultraviolet because in that light we can find differences between the supernova produced by two white dwarfs merging or by a greedy single white dwarf. Just being able to fully model these objects will make our work in understanding the distant Universe and the evolution of the cosmos a whole easier. It is thanks to Type Ia supernovae that we know the Universe is expanding with an acceleration. When observatories like Roman come online in 2027, they will deliver new detections of these events from the far Universe. If we can model them here, we can then explain them there.

'We really need to understand how the UV emission of supernovae behaves locally in order to understand how they behave early on in the Universe. That's the stuff that I've been working on and that's what I've been really interested in,' Dr Devarakonda told me. But that is not all that he does. He works as a liaison between the US federal government and the American Astronomical Society, coordinating funding among the big players of space science in the States, although the latest administration has shown how little it cares about investment in the sciences. Hopefully, at some point he'll get to convince someone to throw some cents towards ultraviolet astronomy.

I am being facetious. It makes perfect sense that ultraviolet observatories are few and far between, with instruments usually attached to space telescopes with wider ranges or broader scopes. Still, the importance that this portion of the electromagnetic spectrum has is undeniable if we want to understand everything in the Universe. Our grip on stars and galaxies goes through our knowledge of the UV. We have seen that individual stars have an impact, some more far-reaching than others. They make even greater changes together. The combinations of all these stars, of all this bright light that we can't see, all these properties that we wouldn't know about at first glance, not only change their local environments, they change their galaxies, and they have changed the whole Universe. We need that kind of knowledge to appreciate their impact, to grasp their immensity and complexity. NASA has an upcoming mission called the UVEX (UltraViolet EXplorer), expected to launch in 2030 as the agency's next Astrophysics Part of its goal is to study cosmic explosions in UV light.[9]

In this chapter we had to quickly adapt to the sudden intensity and brightness of a Universe that was cool and dim. The light might have felt intense, but a new picture is forming. A connection across the wavelengths hints at a deeper truth, and it feels that it is only just the beginning. We know the dangers and risks of ultraviolet light, we have seen the majesty of stars, supernovae and galaxies. We are now on the verge of light as we have never experienced it before. We are about to encounter incredible energetic events and impossible temperatures in the X-rays chapter. We are about to see who truly holds the power in the Universe.

— CHAPTER 5 —

X-RAYS

More Things in Heaven and Earth

> With how many things are we upon the brink of becoming acquainted, if cowardice or carelessness did not restrain our inquiries?
>
> —*Mary Shelley*

> 'Look on my works, ye Mighty, and despair!'
> Nothing beside remains. Round the decay
> Of that colossal Wreck, boundless and bare
> The lone and level sands stretch far away.
>
> —*Percy Bysshe Shelley*

I am something of an eclipse hunter, and seeing a total solar eclipse is one of those celestial spectacles that I would wholeheartedly recommend to everyone. The last one I saw was on 8 April 2024 – my husband and I were in Mazatlán, on the west coast of Mexico. We were on a public beach with tens of thousands of other people. As the Moon covered the Sun, the whole place went silent for a moment before erupting into joyous screams. It felt akin to a religious experience. With the Moon blocking the Sun, the corona became visible. (Not the entire corona, as most of its light is in X-rays, but the part that humanity has been seeing with every eclipse since antiquity.) It was the scientific observations of eclipses that led to the realisation that something very weird was happening around the Sun. The spectrum of the solar corona taken during an eclipse had a spectral line in it that was not known to the scientists at the time. A spectral line is what happens when an atom absorbs some specific wavelength because the light kicks one of its electrons to a higher region or away. For example, if you take a prism and break the light of the Sun, you will see dark lines within the rainbow. Those are called absorption lines. Chemists have catalogued a lot of them, so finding a new one was exciting.

'Somebody worked out that it was probably actually iron, but it could only happen if the gas that was doing the emitting, which is to say the corona, was at least a million degrees in temperature,' Dr Williams explained to me. 'And this was a ridiculous idea at the time, because everybody knew from basic thermodynamics that the Sun's surface was about (depending on where you want to measure it) 5,880 Kelvin. So this idea that there's something a million degrees, so hundreds of times hotter than the surface, in the atmosphere above just didn't make any sense.'

Just like ultraviolet observations, X-rays cannot be carried out from the ground. 'Thank you, atmosphere, for not letting harmful radiation in', some X-ray astronomer might grudgingly say.

So as soon as we have something that could go beyond the atmosphere, we have used it to look at the Sun in X-rays. The first rocket to measure the solar corona was in 1948. Since then, we have been searching and searching for a solution, and it seems we are getting close to one... It is not called the coronal heating paradox for nothing. As Homer Simpson correctly said: 'in this house, we obey the laws of thermodynamics.'* The atmosphere of the Sun cannot magically be hundreds of times hotter than the surface below. It is actually something even more esoteric and mysterious than magic: it is magnetism. The latest theory on how the solar corona gets this hot has come from the incredible observations from two spacecraft: the European Space Agency's Solar Orbiter and NASA's Parker Solar Probe. The two celestial vehicles work alone and together (NASA has contributed to Solar Orbiter) – Solar Orbiter studies the Sun from a distance, and it is the first spacecraft ever to have seen the poles of the Sun. Parker Solar Probe, the fastest ever spacecraft, travels deep into the corona and has been the closest to the Sun.[1] The role played by Parker and Solar Orbiter in understanding the corona is revolutionary. Recent observations suggest that high-frequency magnetic waves are moving through the corona.[2] Models of these oscillations suggest that they deliver enough energy to the atmosphere of the Sun to produce the heating observed. Scientists are not quite ready to call the mystery solved, but it seems to me that they are hopefully on the right track. Solar Orbiter is looking at the Sun in extreme ultraviolet and X-rays, to monitor those oscillations and more in the solar corona.

'The magic thing about the eclipse is you cover up the thing which is really bright in the middle and you get the wispy corona on the outside, which is a million times less bright in visible light. For us, the Sun washes out all that beautiful coronal emission. But if you look in ultraviolet and X-ray light, the Sun is massively different. It is so completely different-looking and it

* During lockdown 2 in autumn 2020, I cross-stitched the quote, as my husband kept coming up with perpetual motion machines he was certain would work.

is beautiful,' Dr Willams very poetically explained. And it is. Beauty is not just in the eye of the beholder. It is also in the part of the electromagnetic spectrum you are looking at.

As Above, So It Glows Below

X-rays are crucial to our understanding of some of the most energetic objects in the Universe – from supermassive black holes to supernovae. (If it's got a super in its name, it probably has something to do with X-rays. Even Superman has X-ray vision!) There are X-rays aplenty around planets, and even on them. The cause of these X-rays is to be found in our perplexing friend magnetism and how it can accelerate electrically charged particles to extremely high speed. Put simply: when those particles interact with others, they create X-rays. Good news for planetary scientists out there. As we discussed in the UV chapter, Jupiter has an enormous magnetic field. Remember it is 20,000 times stronger that of Earth's own, so it is capable of accelerating particles to really high energies. It creates these aurorae mentioned in the Ultraviolet chapter, and as those particles rain down on its large moons, they excite atoms on the ground, which make them shine in high-energy light as well. As above, so below.

Dr Wibisono explained:

> Ultraviolet light and X-rays are very useful in planetary science because they can tell us what a planet's atmosphere and surfaces are made of. The atmospheres of planets absorb, reflect and fluoresce different wavelengths, depending on their composition.

> We see X-rays from the surfaces of Io, Ganymede and Callisto, the Io Plasma Torus, which is a ring of ions and electrons around Jupiter located at Io's orbit, and the inner radiation belts of Jupiter.

Jupiter interacts with the X-rays from the Sun. In particular, researchers saw that, when it comes to lower-energy (also called soft) X-rays, Jupiter appeared to scatter more X-rays when the Sun is more active during its cycle and emitting more X-rays, but it is not as good at doing that for the higher-energy (aka hard) X-rays. Jupiter itself is a source of both soft and hard X-rays in the form of its aurorae. There is a rain of electrons falling onto the planet that are being slowed down by the atmosphere. This is known as Bremsstrahlung radiation, which is one of those wonderfully descriptive German words: the verb *bremsen* means 'to brake' and the noun *Strahlung* means 'radiation'. A charged particle decelerating or accelerating will release light. And Bremsstrahlung's light is X-rays. Those electrons produce the hard X-ray emission from Jupiter.

The soft X-ray aurora comes from different interactions in the Jovian atmosphere. The magnetic field of Jupiter is so powerful that its magnetic field extends for millions of kilometres from the planet. It is the largest structure within the solar system.[3] For reference, the Moon is around 380,000 kilometres away. This boundary is slammed by the solar wind, compressing it. And the compression triggers ions (atoms or molecules missing one or more electrons) trapped in the magnetosphere. The heated ions trigger a phenomenon called electromagnetic ion cyclotron (EMIC) waves. Just like a surfboarder, some of the ions can surf these waves and travel towards the planet. Millions of kilometres later, they slam into Jupiter's atmosphere, interacting with the hydrogen in it and releasing soft-X-ray aurorae.

Dr Wibisono said to me:

> I found that the soft X-ray aurora at the two poles can pulse at the same time, with the same period. No one had seen that before. I also found that the hard X-ray aurora brighten when the ultraviolet aurora show localised brightenings called 'dawn storms' but that the soft X-rays didn't respond in the same way, suggesting that the electrons and ions are accelerated into Jupiter's poles independently.

How cool is that? By monitoring the soft X-ray aurorae in particular, astronomers can tell that the strength of the magnetic field around Jupiter is such that ions travelling towards the planet are stripped of most of the electrons they possess. It can also tell us what kind of elements are out there, such as the volcanic releases from Io that are trapped in the Jupiter's magnetic field.

'Studying the energies of these X-rays can tell us what species these ions are because different ions emit photons of very particular frequencies. We find that a lot of oxygen ions precipitate and emit X-rays at Jupiter's poles. We tend to also find sulphur ions which suggests that these ions originated from Io's volcanoes instead of the solar wind, which was something else I worked on,' Dr Wibisono told me, adding to their extensive list of the coolest observations in the history of astronomy.

X-ray aurorae might be a common occurrence across the Universe, but as far as evidence of that goes, we only have two examples. Jupiter and Earth. Saturn, Uranus and Neptune have stronger magnetic fields than Earth, but we have not seen these X-ray aurorae yet, though they are probably there. So how come Earth gets them? Well, our magnetic field is something to be proud of. First of all, it protects us from a lot of radiation, and it is quite strong, as planets go. Venus and Mars do not have one. The magnetic field around Earth is a shield, but it doesn't just rebuff cosmic rays and the stream of particles from the Sun. Some of these particles can get trapped in doughnut-shaped regions around the planet. These are known as the Van Allen belts. Their discovery goes back to the 1950s and their shape depends on the energy of the particles that get trapped in them.[4] These belts can carry electrons along the magnetic-field lines and send them flying through the atmosphere. There, those electrons slow down. And you know what that means: it's Bremsstrahlung time, baby!

Have you noticed which planet has escaped our scrutiny with regard to aurorae both in ultraviolet and X-rays? It is Mercury. The smallest planet in the solar system has a weak magnetic field and a tenuous atmosphere, not exactly a winning combination when it comes to auroral display. However, researchers believe we shouldn't discount Mercury just yet... At the time of writing, there is a double mission going towards Mercury. *BepiColombo* is a joint project of the European Space Agency and the Japanese Space Agency, two orbiters travelling on the same spacecraft. Getting to Mercury and in orbit around it is not easy. That's because when you launch something from Earth, a spacecraft has its own velocity *plus* the velocity of Earth around the Sun. So *BepiColombo* has gone around Earth, Venus and Mercury multiple times, using the planets to slow down. During a flyby of Mercury, it conducted some preliminary instrument tests and saw something really exciting: an X-ray glow.[5] The mission team believes this to be an aurora-like activity. Electrons are raining on Mercury and pass straight through the thin atmosphere, hitting the rocks below, making them glow in X-rays. The mission team believe that this X-ray fluorescence is an aurora phenomenon and, if this is further proved by observations, once *BepiColombo* begins its science mission in early 2027, it might mean that aurorae of different types are common everywhere in the solar system, in worlds big and small, with or without magnetic fields. You just need something ready to shine interacting with plasma. In some cases, high in the atmosphere. In others, down on the very ground.

A Flash in the Pan

It's incredible what we can study with X-rays. But it is not easy. One challenge is technical: it is difficult to measure the X-ray photons.

'The hardest thing about observing X-rays is we have to do it from space. The X-rays will get absorbed by our atmosphere, which is actually a good thing, otherwise, we wouldn't be here,' Dr Murray Brightman (he/him) from CalTech told me. 'The second thing that is difficult about X-rays is focusing them because they tend to just go through everything. It's very difficult to make a nice X-ray image. So often when you look at an X-ray image, it's just a bunch of dots, and they're not very pretty dots,' he continued.

To turn those not-so-pretty dots into something visually interesting and scientifically useful, scientists and engineers had to come up with ingenious ways of measuring them in a meaningful way. Several observatories now exist in space that can do those kinds of observations. We are not getting the kinds of images we are used to in visible, infrared, or even ultraviolet, but we get some very important insights and we can take a census of what's out there. As it stands, we have measured more 900,000 known unique X-ray sources.[6] Of those, 710,000 sources are supermassive black holes in other galaxies and 180,000 are X-ray-emitting stars from the Milky Way. The remaining 12,000 clusters of galaxies, plus a small numbers of weird objects like X-ray-emitting binary stars, supernova remnants, pulsars, and others. The numbers alone tell of the difficulties we were discussing. There are deep field images from JWST with 45,000 objects in a single photo. It is also important to stress that the census is focused on the consistent sources of X-rays, but there are plenty of sources that only temporarily produce this kind of light, and Dr Brightman is using NASA's X-ray space telescope NuSTAR to study them.

'These are like X-ray sources that appear out of nowhere. There wasn't an X-ray source there before. And we go and look at that region of space again, and we find a new source there!' We have seen that X-ray emission is common across celestial objects, but to suddenly appear out of nowhere and be spotted by our telescopes, they definitely need to have some oomph. 'These are the kind of things that produce X-rays normally, like black holes and neutron stars. Among the sources we find are supermassive black holes eating a star, which is called a tidal disruption event,' he told me.

Tidal disruption events, or TDEs, are great not only because they provide some important insights into black holes but also because they let us explain just how wonderfully dramatic and terrifying black holes are. While some physicists like to think of black holes as celestial plugholes, I think that analogy makes them a bit too active. Most of the time black holes are not eating and are very quiet. Sure, we are going around Sagittarius A* but it's because it is at the centre of the Milky Way. We are not being sucked in from 26,000 light-years away. There are stars getting pretty close – one orbits it every four years[7] – and they are still all right, not running down that cosmic drain. Still, that doesn't mean you should get close. Let's picture ourselves approaching a black hole. The gravity around it becomes so intense past a certain point that even a person will start to feel a different gravitational pull through the body. Your head and feet experience roughly the same pull on Earth but, assuming you're jumping in feet first, the gravity on your lower extremities from the black hole would be different from the gravity in your head. And as you get closer and closer to the black hole, you will start getting stretched. This is a process called spaghettification. Spaghettification means that your body is being stretched. So a star getting too close would be stretched so much that eventually the chemical bonds would break apart until it was just a glowing stream of plasma around the black hole, glowing so hot that you emit X-rays for a short time. That's how TDEs appear in X-rays.

Another fun fact for you all is that if you were falling towards a black hole, you should pick a supermassive black hole rather than a stellar-sized one. The reason for that is to be found in how the tidal forces act. Sure, they depend on the mass of the object but also on the distance and geometry. A stellar-mass black hole is a lot smaller, so these forces will be a lot more violent, whereas picking a massive black hole will give you more time to live before you succumb to the spaghettification fate (which you will do eventually). Astronomers have described a supermassive black hole shredding a star with its enormous tidal forces as 'outrageous'. This is not a sci-fi show – we can't just destroy stars; and yet black holes do it. It is one of the handful of ways that lets us know that supermassive black holes exist at the centre of galaxies. We can measure their properties from the X-ray light produced around them.

At the Edge of the Abyss

Those tidal disruption events might be a quick release of X-rays from black holes, but, as we have seen, there are plenty of supermassive black holes that are shining across the Universe. We can see more than 700,000 of them. They shine because they are feeding. While a tidal disruption event is a quick snack, a little morsel of star stuff, many supermassive black holes are having buffet meals. When these objects are feeding, they are commonly referred to as active galactic nuclei, or AGNs, and X-rays allow us to get a front-row seat for those spectacles. More importantly, they let us pry into some of the crucial secrets of black holes. We wouldn't know all that goes on around them and their wider impact across galaxies without this light.

Dr Dan Wilkins (he/him) from the Kavli Institute for Particle Astrophysics and Cosmology at Stanford University helped me better understand how we use X-rays to study black holes.

Dr Wilkins' work is all about the material getting eaten and how X-rays reflect off it in the final moments before disappearing inside the black hole. His goal is to build a three-dimensional model of what goes on in those extreme environments.

'X-rays tell us about the most energetic processes that are going on out there. In particular, everything that is powered by the force of gravity. It's through X-rays that we can see the behaviour of black holes, that we can actually see material falling into black holes, and that we can see the very close environment around a black hole,' Dr Wilkins explained during our conversation.

Astronomers have been very clever in using the limitations of the field to their advantage. When you get the X-ray emission from a distant source, you don't need to work out what its origin is.

It won't be a star or a planetary aurora. It's going to be the largest source of X-rays in a galaxy: its AGN. Some of those X-rays will come directly from the corona. It's the same word as the one we use for the Sun; this is the region just outside the event horizon of the black hole. That's the closest light from the black hole that can get out. Together with Dr Wilkins, I had another tour guide on our journey around supermassive black holes, Dr Keir Birchall (they/them) from the University of Leicester. The tour was not literal – how cool would that be?! – but necessary. Dr Birchall's particular expertise came in handy when discussing the detailed behaviour of black holes and why we need to study regions such as their corona.

'We can't see black holes, so we need to study, basically, everything else that's going on around them,' they explained. 'Very close to them and orbiting around them there's an accretion disc, just gas and dust moving at, you know, very high fractions of the speed of light. It's very densely packed. There's a lot of friction going on and, of course, that causes a lot of heat to build up!'

They paint a very vivid picture of what goes on there. The corona is shining, the accretion disc is shining, and all this light is bright. The interactions between the coronal emission, the emission from the accretion disc, and the reflection of the light from the accretion disc onto the corona create an everchanging spectacle of high-energy light. I really want to stress that even though the signal recorded is those not-so-pretty dots, researchers can see their changes. Changes in the corona echo across space as the X-rays are reflected off the accretion disc, and due to the limit on the speed of light, those changes act as proxy to understanding the accretion disc, the corona and the whole supermassive black hole environment. The team gets a rainbow of X-rays, maybe not as pretty as the one we see on Earth or that we might measure on Titan, but with an unmatchable wealth of insight into supermassive black holes and their surroundings.

'Putting all of this information together, we can actually build up a map of that extreme environment that's just outside the black hole's event horizon. We can start to see where this accretion disc is located. We can start to see how the corona is connected to it,' Dr Wilkins continued. Combining these maps with the observations from the Event Horizon Telescope we saw in the radio waves chapter has massively improved our understanding. X-ray astronomers apply the radio properties of Sagittarius A* and M87* to many other supermassive black holes.

'Using all of this information we're getting out of the reflection and the echoes of X-rays off the gas, we can reconstruct similar maps of many more black holes across the Universe. We can even measure the fundamental properties of the black hole itself. We can measure the mass of the black hole, and we can even measure how fast the black hole is spinning; we can use this to reconstruct the whole environment around the black hole,' Dr Wilkins explained, before becoming philosophical.

None of us is an island. We do not exist outside our environment or outside time and space. We shape our environments and they shape us in return. Black holes are not inescapable plugholes: remember that they are messy eaters, spewing part of their food out back into their galaxies and even further still. And X-rays tell us that about black holes.

'We can understand how the gas behaves when falling into the black hole; we can understand how stuff gets launched around the black hole. We can start to understand where jets come from,' Dr Wilkins stated.

In particular, when it comes to AGNs, astronomers are seeing the objects in context. The jet of materials and hot gas outflows extend for incredible distances. They carry energy into the galaxy and into intergalactic space. There is a connection between the size of the central black holes and the dynamic properties of the galaxy. The evolution of the two is clearly, intimately connected. It is these observations that are so important to help us make sense of it all. We know so little of their underlying principles, but with X-rays and the information they provide about black holes we are starting to know more.

'I like to describe them in this manner. Your first impression of a black hole is that they're these giant monsters just destroying everything that comes near them. But in some ways, you can think about the black hole as being the sculptor that's guiding the formation of its galaxy,' Dr Wilkins explained in a wonderful analogy before continuing,

> One of the most remarkable discoveries, I guess, over the last probably twenty or thirty years is that the black holes aren't just these ornaments in the sky. If we want to understand how galaxies form, how the structure of our Universe formed, we really have to understand the role that the black holes in the centres of galaxies are playing!

Their impact is not only in the 'intimate' environment in the core of galaxies. It's the whole galaxy. And it's even the groups and clusters of galaxies, where maybe hundreds, if not more, of these objects created the hot gas that rests in intergalactic space.

Dr Birchall explained further how important X-rays are to understand these vast aggregations of galaxies. '[Astronomers] have modelled the emission from the hot gas you find in galaxy clusters. You get lots of galaxies interacting, getting very close together, the gas temperature starts to really, really heat up and so it emits light in the X-rays!'

The hot gas is important in and of itself. The gas is at temperatures of millions of degrees, caused by powerful light hitting matter, a process known as shock heating. Supernova explosions, the jets released by supermassive black holes, acceleration through magnetic fields, and gravitational interactions from the gravity of galaxies moving through it might all play a role in reaching incredible temperatures.

Dr Birchall explained,

> There was not one single point in time when this happened. Some portion of the warm-hot intergalactic medium is composed of particles left behind in these regions when structures that could ionise particles first started to form in the very early Universe. The rest are believed to have been ejected from galaxies at various points throughout the history of the Universe. Given that the intergalactic medium has such a low density (1–10 particles per cubic metre), there is little chance for physical interaction between particles, which limits the pathways by which cooling can take place. Thus, it forms an effective thermal equilibrium at these very high temperatures.

Having more particles would allow the gas to cool down, but instead it is stuck being very hot and tenuous. This intergalactic X-ray emission has also been used in the search for dark matter, or at the very least to say that dark matter is indeed required in the Universe. The Bullet Cluster is evidence that two galaxy clusters collided long ago. The X-ray observations from NASA's X-ray telescope Chandra – another great observatory together with Hubble and the defunct Spitzer – provide a view of where the hot intergalactic gas is located, tracing how the regular matter is interacting in this collision. After all, the gas, galaxies and the light of one cluster will interact with the light, galaxy and gas of the other, and vice versa. But it is also possible to estimate the distribution of matter in the cluster by looking at how it deforms the light of background galaxies. This gravity map doesn't match at all with the distribution of the gas; if dark matter is real, it doesn't have to. Dark matter doesn't interact with itself or regular matter (or very rarely does); it doesn't get affected by light and other forces outside gravity. So it would fly right through the other cluster, creating a much wider distribution than the tight ball of the collision. Just as we see it. Well, we do *not* see it. But that's the whole point: we are finding out things we didn't even know we were missing.

Brighter Is Not a Matter of Size

Several of the big X-ray emitters are the supermassive black holes that actively feed at the centre of countless galaxies. But that is not a fundamental requirement. Some of the brightest sources of X-rays are so small they would fit within a city. They are a special class of X-ray sources known as X-ray binaries. They could be black holes just a handful of times the mass of our Sun or a neutron star, orbited by a companion. Neutron stars and stellar black holes are referred to as compact objects due to their size. For neutron stars, it is estimated that their radius is between ten and fourteen kilometers.In the case of the X-ray binaries, the compact object is a thief. It is stealing material from the companion, accumulating in a disc or on its surface, and releasing a lot of energy in the process. It is staggering how much energy comes out from this process. You'd get tens of times more energy from this process than from the nuclear fusion of hydrogen, and nuclear fusion of hydrogen is what powers most stars. Among these objects there are binaries that can compete in X-ray brightness with supermassive black holes. Their emission is much more energetic than the supermassive black hole feeding process, but their detailed behaviour or how they can be so bright is not clear. They are a juicy mystery for us to sink our teeth into.

'We call them ultraluminous X-ray sources. These are very powerful X-ray sources in galaxies which are not associated with the supermassive black hole in the centre. They're usually outside the centre of the galaxies, but they're more luminous than any X-ray source in our galaxy,' Dr Brightman explained.

Astronomers estimate that there is one ultraluminous X-ray source per galaxy. The Milky Way might not have any, but we have seen galaxies with multiples. Explaining their existence is not easy with our current knowledge. One working theory suggests that these objects release their most extreme light only within a narrowish beam – so they appear extremely bright but, globally, they are not. This could explain most of the sources but not the most luminous. This might also explain microquasar SS 433, which is located in the Milky Way and is very bright in X-ray but not exactly ultraluminous, despite intriguing properties. We might simply be seeing it from the side.[8]

An alternative explanation is that the black hole in these systems is not exactly stellar size, but part of the mysterious class known as intermediate-mass black holes. They go from 100 to 100,000 times the mass of our Sun, and only a dozen likely objects are known. If the compact object is heavier, it means more gravity to generate more X-rays as material falls in. Some ultraluminous sources are probably powered by these heavy-but-not-supermassive black holes. Even if we include this class and the beamed emission, we do not have a full explanation for all the known objects.

'One of NuSTAR's most famous discoveries is that some of these sources are actually powered by neutron stars. This came as a big surprise because they're much more luminous than any neutron star that we've seen previously. So there's been a lot of debate about how they can produce such luminous X-rays. We think it has to do with their magnetic fields,' Dr Brightman continued.

Remember that pulsars and magnetars, particular subclasses of neutron stars, possess stupendous magnetic fields: they might be behind the repeating fast-radio bursts as well as some ultraluminous X-ray sources. Astronomers are looking at both of these classes, a combination of the two, or something else to understand the ultraluminous X-ray sources powered by neutron stars. Dr Alicia Rouco Escorial (she/her), a research fellow with the European Space Agency, told me about this quest to understand the role of neutron stars as X-ray sources, and even took a peek inside neutron stars:

> We have pulsars detected in radio but we also discovered that we have pulsations in X-rays, let's say for ultraluminous X-rays sources. We know now that some of them are neutron stars because we found pulsations in X-rays. The sources are so far from us [that we can't see details so without pulsations] they

> could only be a black hole otherwise. Of course, black holes are very nice, but I'm always trying to tell people neutrons are also very cool.

Based on what she told me, it is easy to agree that neutron stars are very cool. The intriguing object discovered by NuSTAR is M82 X-2. This is an X-ray pulsar that shines 100 times more brightly than an object of its mass should. The magnetic field might be a crucial component throughout the process, but its exact role is nebulous, at best. Dr Rouco Escorial's work focuses on the cooling of neutron stars after X-ray emission. The object accretes material from a companion on its surface. Beyond a certain point, the material experiences a thermonuclear runaway and releases X-rays. This is then followed by a cooling period.

'From that emission you can learn about the structure of the outer layers of the neutron star, but there is so much we don't know.' Dr Rouco Escorial went on to explain the crucial collaboration with theorists and the challenges, like, for example, 'The idea is that we don't know how the magnetic field affects this cooling behaviour!' before admitting, 'We try to avoid magnetic fields because they are always a little painful to put it in a model.' Most physicists would agree with the Insane Clown Posse about the peculiar nature of magnets exemplified in the lyric 'f#$king magnets, how do they work?'. But imagine your magnet is an impossibly dense object, stealing material from a companion star, and releasing more energy than its mass should allow.

Neutron stars are incredible, and, just like black holes and spaghettification, they, too, have a cosmic pasta connection. We have mentioned how these stellar objects are so dense: a teaspoon of neutron star material is as heavy as a mountain. But that is looking at neutron star characteristics in Earth terms. Let's flip that on its head. For humans, Earth is quite rugged. Mountains and ocean trenches are extreme, but it goes up and down a fair bit. But taking the planet as a whole, it is smoother than a billiard ball. Let's embrace that. Holding the world in the palm of your hand feels relatively smooth... and then you try it with a neutron star. The smoothness is inconceivable. A neutron star is thousands of times smaller than our planet and its 'mountains' are about a few millimetres tall. That means that it is a billion times smoother. Underneath the oh-so-perfectly-smooth surface there is the crust of the neutron star, 10 billion times stronger than steel. This crust is believed to be organised in nuclear pasta. As mentioned, there are a lot of neutrons in neutron stars but also protons, and their interactions in this extreme environment create distribution of matter in shapes that might look like gnocchi, lasagne, spaghetti. I'd argue that they are not spaghetti per se (they are bigoli as they are thicker) but to be fair, among the nuclear pasta there are waffles as well as anti-spaghetti and anti-gnocchi. Gnocchi are little clumps, anti-gnocchi are filled blocks with little holes. If this is not already cool enough, Rami Mandow (from the radio waves chapter) told me that there might be tiny quantum tornadoes going through the layers of nuclear pasta. If two tornadoes hit each other, they can change the speed of the whole neutron star. They call this a 'glitch'. You would think that neutron stars with their extreme energy, quantum tornadoes and pasta-shaped filling get the crown for being the weirdest X-ray sources, but the most bizarre is still to come.

You Can't Sit With Us

There's a lot to talk about neutron stars in ultraluminous X-ray sources. As we have seen, they are brilliant, in every sense of the word. But there has also been a weird population featuring black holes and this has taken a long time to become accepted as a prominent class of X-ray binary sources. These black holes are found in globular clusters, spherical collections of tens of thousands to millions of stars that orbit galaxies. It was assumed that black holes in these objects could not exist. The first calculations, over fifty years ago, showed that they would sink in the centre and their gravity would make stars go faster and faster until the cluster dissipated. In the mid-nineties, new calculations suggested the opposite. The black hole would fall into the centre but the interaction would send it slingshotting out. It took another decade for calculations that showed that black holes might be present in these globular clusters after all. I spoke to Dr Tana Joseph (she/her), an astronomer from South Africa, who was part of the team discovering this population and its peculiarities.

'Then it was like, actually, we might find some, but they'll be rare. Black holes in global clusters will be rare. Do you know how we count in astronomy? If you have zero, it's rare. If you have one, then it's not ubiquitous, but it's normalised. If you have two, then you have a population,' Dr Joseph explained.

The 'population target' was reached as she joined the team in 2010: they found the second black hole in a globular cluster. Today, there are twenty black holes known in globular clusters and they are ultraluminous X-ray sources. That's about 1 per cent of all known ultraluminous X-ray sources. And in a field where there's a lot of uncertainty and many assumptions on how extreme objects produce extreme light, this weirdness pushed them to the side.

Dr Joseph continued:

> In 2011, there was a review of ultraluminous X-ray sources. And our sources, these black hole binaries in globular classes, were not mentioned because there were so few of them. And then a few years later there was another review, still no mention. Then there was another review a few years after that and there's a paragraph in a whole review paper, there's one paragraph that acknowledges the existence of these things. And now we start to get to the point where there's enough of them that we can't refute that they exist.

The team believes that when the space-based gravitational wave observatory LISA (Laser Interferometer Space Antenna) launches in the next decade, scientists will have the confirmation they need to prove that the rare X-ray sources involve a black hole. But these sources are small and far away and there is no direct evidence yet. LISA will detect gravitational waves from sources such as these binaries.

> We are deducing that they are black holes due to circumstantial evidence. We do not have mass estimates of the components because you basically can only see them as a single ultraluminous X-ray source. You gather the evidence to show that it's extremely unlikely that this is a neutron star. You need to 'Occam's Razor' it [Dr Joseph jokes]. Is this twenty neutron stars in a trench coat, or is it one black hole? You look at data that you have and you build up the case that these things are probably black holes.

The reason why LISA is needed and ultraluminous X-ray sources cannot be found with current observatories is that researchers believe that the companion of these black holes might also be a weird object: a white dwarf. They expect an enormous number of these compact binaries to be found in the Milky Way and its surroundings, so it would give an idea of just how many we would see in globular clusters. These peculiar sources will finally have a seat at the table.

It was an interesting discussion to have with Dr Joseph. The sources are referred to as ultraluminous because the limit for their luminosity is defined by the Eddington luminosity – basically the maximum a star can achieve if its radiative power needs to be balanced against the gravitational pull. A measurement that feels a bit out of place when it comes to objects that routinely defy that limit, and it's especially galling since Eddington went to great lengths to publicly attack and mock Subrahmanyan Chandrasekhar for his theory on what happens at the end of the life of stars, and in particular the Chandrasekhar limit we encountered when discussing white dwarfs in the previous chapter. Eddington's ability to push the scientific community against Chandrasekhar kept scientists from fully exploring the possibility of black holes for years.*

'Since its beginning in the 1960s, X-ray astronomy really has just been surprise after surprise after surprise,' Dr Wilkins told me. And he continued: 'One of the most remarkable results of

* Chandrasekhar is reported as saying that, despite Eddington's behaviour, he continued to admire him and consider him a friend. Several times I was also told when writing this chapter that Eddington might have been gay. I couldn't find any evidence beyond him being a lifelong bachelor, but I include this here because queer people, too, need to remember that our forebears might have been absolute dicks! Isaac Newton, another confirmed bachelor whose rudeness is remembered alongside his achievements, might also feature in this category.

these first rocket experiments with just a little X-ray detector is that what we previously thought was empty space is brighter in X-rays than the Sun, which led to the question: where are all these X-rays coming from in the Universe?'

The Sun emits X-rays, as we have seen, and those rockets were sent up to study the corona, but most of the X-rays detected were coming from the myriad black holes found around the cosmos, and it was X-ray astronomy in the 1970s that brought the observational evidence that black holes were real.

Dr Wilkins continued:

> Before this, black holes were a piece of theory. They were predicted by Einstein's theory of general relativity. Some of the most famous scientists we associated with black holes thought that these things surely can't exist. The theory of them is just too weird. This includes Einstein himself and Stephen Hawking, another famous name we associate with studying black holes. In those early days, they both thought that the Universe can't possibly allow these things to exist. They're much too weird. Then X-ray astronomers discovered this bright, rapidly flickering X-ray source in the constellation of Cygnus.

That object is now known as Cygnus X-1, and it is an X-ray binary, a stellar-sized black hole stealing plasma from a supergiant star.

In X-ray astronomy we see a running contrast that can easily be applied to the rest of astronomy, science, or society: an attachment to simplicity, assumptions, to neat little boxes that perfectly explain what we see. It's where we use the Eddington limit, which is built on so many limited assumptions and we keep finding objects that break it. It's thinking that collapsing stars can't turn into black holes, that black holes can't exist in globular clusters, that something else is producing X-rays at the centre of galaxies. And on the other side is reality: messy and complex, where the neat little boxes crack at the seams long before you can put a lid on them. Modelling and simplifications are important stepping stones, both in science and in society, but we must never forget that they are coarse approximations. Just because modelling magnetic fields is complicated, we cannot pretend they do not exist. And we saw that good scientists don't do that. I find that the beauty of this Universe is not in finding the right box to fit in, but in pushing at the crack in your box, making it more comfortable, or breaking it apart altogether, to where you can even find inspiration from black holes in globular clusters!

— CHAPTER 6 —

GAMMA RAYS

The Pursuit of the Energy of the Extreme

In order to rise from its own ashes,
a phoenix first must burn.

—*Octavia E. Butler*

I hear you whispering there O stars of heaven,
O suns—O grass of graves—
If you do not say anything how can I say anything?

—*Walt Whitman*

And now we have reached the domain of the light of which we have no direct or indirect experience. At least I hope so. Gamma rays are the type of light so powerful and energetic that you need to have many centimetres of material to shield yourself from it. And that's what you get from nuclear reactions. Many decades of concern (justified or not) about nuclear power have instilled in all of us the knowledge of the danger of radiation – yet the energy carried by the gamma photons in nuclear reactions truly pales in comparison to what you get from celestial events. Light thousands of times more powerful is emitted by the common bodies found peppered throughout the Universe. But as their properties become extreme, so do their gamma-ray emission, and eventually we find the most cataclysmic and energetic events. The highest gamma energy ever recorded in space is 2.5 million times the highest energy you get from nuclear decay. But this wasn't even the bulk energy of the event; it was in fact the energy of a single photon.

The energy of a single photon is usually something that can only be compared to microscopic things. But not this gamma-ray photon. The easiest comparison is to consider a mosquito in flight, drifting through its day, not paying attention, and suddenly crashing into you. It might not be very noticeable, but you might sense it. Now, consider twenty careless mosquitos hitting you in the same place in one go. You'd probably feel that. It might be light but it's definitely something you can imagine having a physical impact. Well, that one gamma-ray photon had the same energy. I'm going to say it again: this is the energy of a single photon. A single particle of light. This is trillions of times more energetic than the visible light you are getting while reading this. It is an enormous amount.

The event that sent this our way was a gamma-ray burst: a brief emission of gamma rays from an extreme event in the cosmos. Gamma-ray bursts have been measured since the 1960s and this was by far the strongest. Those record-breaking photons pummelled the Earth's atmosphere for 800 seconds. I am using the word pummel in a very literal sense. The effect of this powerful light actually affected the ionosphere – the ionised portion of the atmosphere – in a similar way that a solar storm does. The atmosphere of our planet was changed by an event not just from beyond the solar system, but from a whole other galaxy, the dramatic effects of an incredible supernova going off about 2.4 billion light-years away. For many hours after the event, called GRB221009A,[1] the Earth's ionosphere continued to be disturbed.

Take a minute to reflect on that statement. I regularly write about the bold and the powerful of the Universe and even I have to take a deep breath after reading about this particular event. Every image we get from space reminds us that we inhabit a wafer-thin portion of the planet we call home. A little volume of its thin atmosphere keeps us safe, warm and alive. And then you discover that events in another galaxy altogether can mess with your little protective bubble. Take a leaf out of the great philosopher Dolly Parton's book: don't let it drive you crazy. Luckily, it is estimated that events such as this only happen once every 10,000 years. And we can't be too scared of them. One way or another, we are all here thanks to the processes that created these brightest flashes.

Starkiller Phase

Gamma-ray bursts (GRBs) are mostly produced in supernovae and in collisions between neutron stars. We have discussed just how powerful and dramatic these events are, but what if they could also offer a helping hand for life? Supernovae and neutron star collisions produce and disseminate so many elements that allow life on Earth to exist as it is today. Supernovae exploding in the solar neighbourhood might dramatically affect life – the light from them could be dangerous (especially the gamma rays) and there can be a stream of charged cosmic rays that could deplete the ozone layer. We do not know if there is a distance where we are safe, but between forty to fifty light-years away we should be safe enough, as long as we do not end up having a jet pointed at us.[2] We have no star risking an explosion in that proximity for the foreseeable future. A closer supernova would affect the way our atmosphere protects us against dangerous light such as ultraviolet; if more UVs are pouring in, we might see an increase in mutations that will affect many species. For some, this would mean an early grave, but other species might instead use that as an opportunity to thrive. Maybe this is exactly what has already happened. Some research argues that supernovae might have influenced the biodiversity of the last 540 million years.[3] Are we children of the stars, then? Phoenixes using not just their ashes, but their death throes to build better offspring? Perhaps. These connections between the celestial and the terrestrial have some interesting ideas and some supportive evidence, but there is a lot that we do not know about both the emergence and divergence of life on Earth and the lives and death of the stars. Again, as above, so below. Though gamma-ray bursts do provide a way of dispelling some of that lack of knowledge – and in spectacular fashion!

'Basically, GRBs are super important because they're some of the most violent explosions in the Universe-and there are two different kinds of GRB populations,' Rahul Jayaraman (he/him) explains. Before moving to Cornell University, Jayaraman was a graduate researcher at the Massachusetts Institute of Technology (MIT) using TESS (Transiting Exoplanet Survey Satellite) to study the emission of gamma-ray bursts in optical wavelengths. We have seen how events and objects are best understood by looking at them across many wavelengths and, of course, the events that are remarkable in gamma rays offer fascinating insights into other wavelengths. There is also another fact that makes multiwavelength observations important here. It is very difficult to observe gamma rays. Gamma-ray astronomers might count each photon they get from a source, and the number of known constant sources is just over 7,000, but when it comes to GRBs they also disappear pretty quickly, so follow-up observations provide more insights into them. But being a fleeting event doesn't mean they don't have an impact. They are the most energetic and luminous electromagnetic events to happen in our Universe since the Big Bang.

'If you have a gamma satellite and there is a gamma burst, for instance, then the GRB would be brighter than the entire sky in gamma rays. Every star, every galaxy in the Universe combined for those few seconds,' Professor Tanmoy Laskar (he/him), from the University of Utah, Salt Lake City, offered me this nugget of information in the most casual way possible as my jaw dropped. The brightest events in the Universe and we can see them! It's incredible; and we found them by chance! 'Before we looked at the Universe in gamma rays, we didn't know that these things even existed. We didn't know that these explosions took place. We didn't know that there could be such energetic events, that the Universe could make these things, and that physics operated like this,' Professor Laskar continued, now truly selling me the awesomeness of GRBs. It was not difficult, to be fair!

The discovery of these events comes from military satellites in the 1970s. Satellites were sent to space to keep an eye on the Earth. Why do we need to look at the Earth in gamma rays? Well, to see nuclear explosions, of course! This is one of the many examples of the overlap between astronomy and the industrial-military complex, which for a pacifist like me is not something I like at all. Personally, I would like the expansion of human knowledge without the accompaniment of increased human suffering. Still, in 1974, satellites that were hunting for nuclear explosions on Earth found flashes of gamma rays beyond our planet. It became clear that these events were so powerful that they could not even be in our galaxy. They came from far beyond it. Over the following five decades, researchers have been able to divide them into two main populations: the long gamma-ray bursts are most probably produced by supernovae and last from a couple of seconds to several minutes. About 70 per cent of GRBs detected are of long duration. The remaining 30 per cent are the short gamma-ray bursts and the mechanism behind them is the collision between neutron stars. The categories have their exceptions. For the short ones, there are four examples found in the last fifty years associated with flares from magnetars, the extremely magnetic neutron stars we mentioned earlier. There are also some events beyond these two, called ultra-long gamma-ray bursts (more than 10,000 seconds) that might be the results of many different highly energetic events, like the destruction of a star by a supermassive black hole, as well as supernova-related events. Professor Laskar explains,

> Whatever the mechanism, some fraction of these events produce very powerful jets. So these jets are travelling at 99.99 per cent of the speed of light. Very close to the speed of light. We don't know how these jets are formed. We think there must be a black hole that is formed either at the centre of the supernova or when two neutron stars merge. But there are other theories as well. If these jets happen to be pointed at us, then you get a burst of gamma radiation from the jet. Again, we also don't know how the gamma rays are produced. But we think that something happens inside the jet that makes the gamma

rays. And then when the jet interacts with the surrounding medium, it produces what is called an afterglow. That shows up in X-ray wavelengths, visible light, and also in radio waves.

Teamwork Makes the Dreamwork

Professor Laskar's words and work stress the crucial importance of multiwavelength astronomy, of looking at things with different light. Each wavelength of light might provide unique insight and new discoveries, but it is by using them *together* that we reach the necessary holistic understanding of a specific phenomenon. Sometimes not even all the possible light is enough, at first, to get the full picture. A perfect case of this is probably the most famous short GRB of all: GW170817. This event was caused by the collision of two neutron stars and generated gravitational waves which were detected by the LIGO (Laser Interferometer Gravitational-Wave Observatory) and Virgo detectors here on Earth – respectively in Washington State, Louisiana and Tuscany on 17 August 2017. This was the first gravitational detection ever of two neutron stars. Just 1.74 seconds after that detection, NASA's Fermi gamma-ray telescope detected a GRB lasting about two seconds. An alert was issued fourteen seconds later. Several groups moved their telescopes in that direction – when it became clearer that the two were connected, more and more telescopes got involved. That meant a lot of people observing and trying to understand what was being observed. From August, when it was detected, until the announcement exactly two months later, this was astronomers' worst-kept secret. Don't get me wrong: they did very well keeping a lid on it officially, but the rumours were so consistent that many outsiders knew what was coming. We could be glib and say that this proves that we definitely have not found aliens, simply based on the fact that scientists won't be able to shut up about such a discovery. But I think it is important we celebrate what a global accomplishment this was. The official announcement paper was co-authored by almost 4,000 astronomers – often reported as about one-third of all working astronomers in the world. They were at 900 institutions and used data from seventy observatories – on all seven continents and in space. What better testament to the positive power of science than its ability to unify people around a common quest? The quest in this case was a neutron star collision which created ripples in space-time and a lot of light. And the cherry on top of this incredible discovery? This short GRB is not quite right!

Imagine waiting for decades to find irrefutable evidence that short GRBs are caused by neutron star collisions and, once you have that evidence, that event doesn't exactly match its model. Even its X-ray afterglow caused perplexity for years after the collision. But gamma rays alone can't deliver all the necessary insights to build a coherent picture of why this event was not as expected. When it comes to transient events in astronomy, two seconds might be long, but if you are trying to build a coherent picture of the phenomenon, it's very short, so you need to observe such an event in as many wavelengths as possible.

Jayaraman explained,

> Some people say that the GRB associated with GW170817 was kind of anomalous because it was a little bit weaker than we would expect. The prevailing hypothesis is that we were looking at it kind of a little bit off-axis. Like, the jet wasn't

> directly pointed at us, we were slightly offset. We got some fraction of the radiation in the jet, but not the entirety of the jet. And so multi-wavelength follow-up is important because the GRBs emit differently in different wavelengths.

Before we jump on the insights we are getting from multi-wavelength follow-ups, there is another obvious approach. We just need to find more of these GRBs, but that's easier said than done. The current estimate is about forty detectable gravitational events involving neutron stars per year. We do not see that number, as the detectors are not on all the time. Imagine if we could: each detection is a treasure trove of possibilities. Several scientists are working to expand this number, among them graduate researcher Madalyn Johnson (she/her), from the University of California, Santa Cruz.

'We're looking at where LIGO is detecting these sources and immediately using a telescope. We're looking at these at the same time to see if we're looking at the same source and trying to make a detection and linking those gravitational waves and gamma rays,' Johnson told me.

Gravitational wave detectors work a little bit like seismometers. To triangulate a source, you need to have detection in at least three of the now four gravitational wave observatories on the planet. As I have already mentioned, they are not always switched on, which is a hurdle in terms of continuous discoveries. The goal is to see more gravitational waves from neutron stars and GRBs, but it would be interesting to see if the collision between a neutron star and a black hole or between two black holes releases light in some shape and form, something that has not yet been witnessed. While observational gravitational astronomy is in its infancy, with the first gravitational wave detection having taken place in September 2015, at least those waves can move through everything in the Universe without being stopped, which is a crucial advantage.

'Electromagnetic waves often get absorbed or scattered before reaching Earth. Light from cosmic events has a lot of dust and particles and all this stuff that it needs to get through to reach us, and a lot of time it can get distorted,' Johnson explained. 'We can increase the number of events we detect by using different methods that don't get distorted or obscured, as electromagnetic waves can.'

But How Do They Work?

We can indirectly follow the power and devastation of gamma rays using instruments that can get a lot more data out. The GRB is the camera flash and we are left following up in the splotches of coloured light that remain in our vision. Obviously the physical process is not the same as our eyes getting too much light; it's the material that interacts with the GRB that gets all the excitement. The excitement has a specific name when we consider the electrons in this material. It's called synchrotron emission. Here, electrons are accelerated along the magnetic field lines. When an electrically charged particle is accelerated, it releases energy. The bigger the acceleration, the bigger the energy. At first you get a big energy release, which comes as an afterglow seen in X-rays. These afterglows are visible for several days after a GRB. Over several weeks and months, these energies will be weaker and the afterglow is usually studied in visible light and radio waves. But don't think that the afterglow seen in our arbitrarily divided range of wavelengths is just a repeat. X-ray observations are not just a preview of optical views and, in the same way, optical glows are not just a tease of radio. They all can capture different moments *as well as* different portions of this event.

'Multiwavelength follow-up is important because the GRBs emit differently in different wavelengths. The initial gamma rays are typically caused by shocks within the jet: that's super-high energy, non-thermal emission. And then you have the afterglow; that happens when the jet hits the circumstellar (circumburst) medium that typically radiates in redder bands,' Jayaraman summarised. He went on to explain that the properties of the afterglow tell us the details of the burst we wouldn't learn from gamma alone. 'We can really constrain a lot of the physical properties from the burst by extending our observations to wavelengths that are not exclusively in the gamma-ray band.'

It's a great thing to be able to say: 'We don't know.' The exceptions to our rules are reminders that our explanation needs improvements. In the case of GRBs, those improvements come from multiwavelength observations. A simple example of these events is known as the optical flash. We have seen thousands of GRBs over the last thirty years and a few them have shown something rather peculiar: when the gamma-ray burst goes off there is also an optical flash associated with it. It is unclear what drives the optical flash. It seems to have a roughly similar size and comes from the same region as the gamma-ray emission, and yet all the models for the energy of this flash, based on standard GRBs are off by a very great deal.

'We're not sure if there's something else coming into play or what's going on there,' Jayaraman tells me. 'That's sort of what I'm trying to look into and investigate. And I'm trying to search for these kind prompt flashes and learn what they are like.'

The components in different wavelengths are not interchangeable. The mechanisms behind the different emissions are not one and the same. They are probably complementary and interconnected, but one type of observation doesn't solve every mystery.

Professor Laskar added:

> There is the process that produces the gamma rays. We don't think that is the same process that produces radio waves. So the gamma rays come from inside the jet. And they last maybe ten seconds, depending on what kind of event it is. Sometimes they last several minutes. Then the jet interacts with the environment and produces this afterglow. That afterglow produces radiation at all the different wavelengths. So it's a rainbow of colours; it's a rainbow across the entire electromagnetic spectrum, from the radio to the X-rays and maybe even gamma rays. And that radiation from the interaction of the jet with the environment, that lasts for a long time, that can last for years.

Invisible rainbows keep popping up in our celestial observations. Studying the afterglow across many types of light tells us so much about the explosion that created it. We can start by learning about the jet, its properties, how fast it was moving. We can learn how much energy there was in the GRB. Certainly, it can help work out the distance. Astronomers can also learn how dense the environment around the explosion was. And from that density they can guess what kind of progenitor we are talking about. Supernovae and neutron stars are not distributed evenly through the heavens. All this is already a testament of what multiwavelength observations can do when it comes to GRBs. But there's more. Because these events push electrons to energies that we can't recreate in our labs, by studying the afterglow we are looking at physics at the edge of our knowledge. We are looking at the highest energies since the Big Bang. Ultimately, we want to try and understand it all. Why does only 1 per cent of supernovae produce such GRBs and how can they create light so powerful that it can shake the atmosphere of a blue planet around an average yellow sun 2 billion light-years away?

The Usual Suspects Return

Gamma-ray bursts don't keep me up at night. If we were in the beam of a GRB, even 200 light-years away, Earth would be vaporised, but luckily there are no stars that might create a GRB within that distance, so we are safe.[4] If anything about the gamma-ray Universe were to keep me up at night it's how little we know. As I said, we know of just over 7,000 sources – just to put the numbers in astronomical context, we know 3,800 comets in the solar system. In general, instruments like NASA's Fermi are not sensitive enough to detect so many sources that should be out there. Our Sun, for example, emits gamma rays. The 7,000 number comes from the Fermi catalogue and in there the most prominent feature is the plane of the Milky Way. It is there, bright and central, telling us that there are sources all around us. Now there is plenty off the plane of the Milky Way, too, and they are the characters that we have become familiar with in the course of the book.

We mentioned at the beginning of the chapter that radioactive decay creates the gamma rays on Earth. That happens in space too. There is also the annihilation of matter and antimatter. Neither are super common. There is little antimatter in the Universe and the elements that decay as gamma rays are also very rare. A major player in the more constant source of gamma rays is inverse Compton scattering. You have a nice photon moving about with its own energy and then *bam!* it collides with an electron that has been accelerated to almost the speed of light. In the collision, the photon steals some energy and suddenly it is a gamma-ray photon. With this in mind, the location for these robberies becomes obvious. You need places of intense magnetic fields and a lot of plasma. These wild things are mostly found around the extreme events.

It is not surprising that most objects found in the Fermi catalogue are active galactic nuclei (AGNs), the very luminous state of accreting supermassive black holes. You've got plenty of high-energy particles and photons, strong magnetic fields and high temperatures. Shake it all together and you have the perfect recipe for gamma-ray production.

'The majority of those are actually AGN, which is good and also bad because, basically, when you try to detect new sources you have to ask yourself, well, is that an AGN or is that something I'm interested in?' joked Dr Yuzhe 'Robert' Song (he/him) as he took me on a tour of the gamma-ray sky beyond gamma-ray bursts.

Dr Song tells me that Yuzhe means 'philosophy of the Universe' and I couldn't not include this beautiful example of nominative determinism. AGNs have popped up across the wavelengths all the way from radio, so it is not surprising that these old friends are here. Gamma observations add another important piece to the puzzle.

Madalyn Johnson explained,

> There's a lot of information that we still need to fully learn about AGN and I think gamma rays are a very essential part of that understanding. They are very energetic events that can emit a lot of high-energy gamma rays. Studying AGNs with gamma rays can tell us how and where these very high-energy photons are created. It can give us a better understanding of the jets released by AGNs, and the source of the extreme luminosity that comes from the AGN. Additionally, we can better understand supermassive black holes and how they are evolved.

Every single point is correct. This argument should be very familiar at this point. Observers in the other wavelengths made similar cases for similar – if not the very same – properties. The difference is in what is being observed. The AGNs in question are the ones with the powerful relativistic jets. This suggests that the gamma-ray emission is not thermal. It's not an incredibly hot surface: it's due to some extremely energetic physical process.[5] Radio observations can also be used to probe these non-thermal events, especially those AGNs releasing energetic jets. But gamma rays take you right to the centre of the action, close to where the jet is being created, and that matters because we do not have a clear idea of how those jets are formed. These supermassive black holes are releasing beams of plasma moving close to the speed of light that can extend for millions of light-years. The combination of radio- and gamma-ray observation are building a coherent picture for their formation and evolution. This means that it is not just the properties of AGNs that are being tested, it is also their behaviours. As we now know, supermassive black holes flare up when a star gets too close and, in some cases, when a star is ripped apart and the stellar entrails are splayed out for everyone in the Universe to see.[6] I mean that literally. This process can sometimes be as bright as all the stars in the galaxy that host that supermassive black hole. Gamma rays provide us with a view from the thick of it.

You do not have to be an active supermassive black hole to get some unique insights, thanks to gamma rays. Sagittarius A*, the supermassive black hole of the Milky Way, is as quiet and as relaxed as they come, but it has had its period of activity, even in the recent past. We know that because there are two enormous features that we can still see linked to it: the Fermi bubbles. A team of astronomers led by Tracy Slatyer from the Massachusetts Institute of Technology was looking for a faint but specific gamma-ray emission, considered possible evidence for the existence of dark matter. Slatyer did not find evidence of dark matter but she did find two colossal bubbles of hot gas extending both above and below the Milky Way's plane. They extend 25,000 light-years from the galaxy's core – a distance similar to that between the solar system and the Milky Way's centre. It is at the very centre that we need to look for an explanation. The leading hypothesis links the formation of the Fermi bubbles with a dramatic flare from Sagittarius A* about 3.5 million years ago. Hot gas was thrown out at millions of kilometres per hour and it remains there, glowing slowly, above and below our galaxy. Two enormous features of the galaxy were completely invisible to us, and only this investigation into gamma rays brought them to light.

There are many sources of gamma rays that are not as explosive or dramatic as the supernovae, or the neutron star collisions, or the supermassive black holes. These sources are varied, and common as well. It is not surprising that the plane of the Milky Way is so bright in gamma rays when we consider all the objects that produce gamma rays. And as long as there is energy to produce them, we will follow them to their sources. Many objects in space can do just that but using a different method from what we have discussed before.

'Pulsars are stable and not explosive. And globular clusters, too, although their gamma emission is possibly a collection of different gamma emission processes that involve pulsars. There are little bubbles off the galactic planes and those are actually star-formation areas. Those are definitely non-explosive,' Dr Song explains.

Once again we are witnessing particle acceleration and interactions as the cause of gamma-ray emission, though slightly less direct. Here, protons are being accelerated. Hydrogen, the most abundant element in the Universe, is just a proton and an electron. Most of the hydrogen has been ionised so the electron was stripped away, leaving you with just a positively charged proton. If you have a strong enough magnetic field, you can bring protons to relativistic speed, very close to the speed of light. Then you smash them into other protons and you have some fun, weird particles produced. Humans have learned to do so in particle accelerators. The magnets at the Large Hadron Collider – part of CERN – are more than 100,000 times stronger than the Earth's magnetic field. We humans are chasing higher and higher energy to study the most extreme conditions – such as the time right after the Big Bang – but you can accelerate protons to interesting speeds with less. Besides, nature easily comes up with stronger magnetic fields out there. Jupiter alone can compete with particle accelerators. Around pulsars and magnetars, collisions between protons can create a short-lived particle called the neutral pion.

We met quarks in the microwave chapter. Quarks don't like to be alone. They are either in pairs or in a throuple. A pair of quarks is a meson. The latter is a hadron (careful with the spelling when you are Googling them). Both protons and neutrons are hadrons. A proton is made of two up quarks and one down quark. Now, when you slam a proton against another proton you get some antimatter out. In particular, you get some antiquarks. These have the same mass as the quarks but opposite electric charge. The collision between two protons creates a situation where an *up* quark can join an *antiup* quark or a *down* quark can join an *antidown* quark. That's a neutral pion. And the cool thing about pions is they last a fraction of a second before just turning into gamma rays.

'The neutral pion is the product of proton-proton interactions. Basically, we have to have relativistic protons hitting another target population of protons. And that can produce the neutral pion that we see decay to gamma rays. And, actually, that emission is quite characteristic because it peaks in an observing sweet spot for NASA's Fermi,' Dr Song told me. 'If you think about it, that's why we see so many gamma rays coming from the galaxy itself, because the galaxy is full of hydrogen gas. And whenever the cosmic rays hit the hydrogen gas, in the gas itself, we see the gamma rays.'

Accelerating these particles still requires a lot of energy, but not the enormous amount seen in gamma-ray bursts. They are found in many different objects, beyond pulsars. The shockwaves created by supernova shells slamming into the matter between stars can do that. It is also possible to get gamma rays from stellar flares. Any stars, not just the big ones. Our Sun and much smaller stars, such as the red dwarf, can create cosmic rays that then go to create gamma rays. But we do not see this emission. At least not yet.

'The same mechanism that produces all those gamma rays should produce gamma rays from the red dwarfs. Red dwarfs are thousands of times more magnetic than the Sun, so they should absolutely emit gamma. It's just that Fermi is not sensitive enough to detect them because they are small and far away,' Dr Song said, hoping for future observatories to be more sensitive.

There are also some white dwarfs that can occasionally release gamma rays as they go through a nova phase, when they steal material from a companion and suddenly become a lot brighter. And maybe these events are a prelude to a bigger event, a supernova explosion when the white dwarf has stolen enough material, or a collision to become those special Type Ia that can help us measure distances in the Universe. Gamma rays could be emitted in those events, as well as gravitational waves that might be spotted by future observatories. There are still many pieces missing in our understanding, especially when we seek to work out how the Universe in gamma ray works.

'Gamma rays can help us understand some of our Universe's most powerful and energetic events like supernovae, active galactic nuclei, black holes, and a lot more. And studying these objects can provide insight into their creation and better shape our understanding of the Universe,' Madalyn Johnson perfectly summarises.

Gamma-ray astronomy is challenged by incredible technical limitations and yet astronomers continue to find ways to get around them. It's a good thing they do. We might have discovered the extreme events and objects that populate the sparse gamma-ray sky with other wavelengths, but we would be missing some crucial understanding about the Universe. I hope I've made it clear that we do not have all the pieces of the puzzle, but thanks to the synergy between gamma rays and other wavelengths, we can at least tell the vague shape on the front of the box. Putting a couple of corner pieces down.

Without gamma rays, there are so many things we couldn't discover and yet you need to use everything to really understand the Universe. This is a perfectly fitting ending for your journey beyond the visible. We have embraced all the wavelengths and at the extreme: it is clear that we need to use all of them. We humans have limitations as individual beings, as a species, and in our technology. Why should we make it difficult for ourselves by ignoring parts of the electromagnetic spectrum that might deliver new insights and show us new phenomena?

We should go further. We need to study the Universe in gamma rays even without the revolutionary realisation of a new phenomenon. We have seen that even mundane sources like our Sun and other stars emit gamma rays. They might be occasional, difficult to spot and to study, and may not be very interesting compared to explosive gamma-ray bursts that can evaporate a planet 200 light-years away. However, we can't be sure until we check (and I bet we will still learn something new). The whole astronomical endeavour has shown that the unremarkable is remarkable when it comes to the Universe.

Humans have learned that there are invisible rainbows across this wonderful Universe and we must seek them out in every nook and cranny. Without all the wavelengths of light, you can only have a partial view of the Universe. And how much could we possibly understand of that portion if we don't try to expand our horizons?

EPILOGUE

Into a Universe Without Shadows

> E quindi uscimmo a riveder le stelle
> (And so we got out to see the stars again).
>
> —*Dante Alighieri*

From the longest wavelength to the shortest, from the most energetic to the weakest, something stops light, and a shadow is cast. At the beginning of this journey, I stated that I didn't want to be considered a priest of science. You have seen that, while I get excited about discoveries, I do not shy away from the murky and often horrible behaviours of science and scientists. Science is human, and that means it is flawed and messy. It is a tool that can deliver the most incredible findings, but also do unspeakable evil, which we must speak about. Otherwise, it does a disservice to us, to science, and to the next generations. Science is also about choices, and we must be conscious of our choices. In writing this book I have made many, but one of them has been the driving force behind most of it, and it needs to be made explicit. Some of you might know already. Some might suspect. Others might be surprised. Every single person interviewed for this book is LGBTQIA+. They are the other invisible rainbows that have for so long been beyond the interest of scientific institutions. They are truly extraordinary researchers in their respective fields. All of their identities are part of who they are and yet if the person is not a cis, straight, white man from the Global North, suddenly the choice of featuring them becomes political, as I was told in the making of this book. Indeed, across all aspects of my work, I hear over and over again that I bring politics into science. The reality, though, is that far-right reactionaries without the ability of a consistent or factual argument enjoy the fact that certain identities are politicised. My own existence as a queer scientist is deemed political. Let's consider that the word 'scientist' itself was politicised because it was coined to move away from the term 'man of science', used previously. William Whewell wrote it in a review of Mary Somerville's *On the Connexion of the Physical Sciences* (1834) that she could not possibly have been a man of science, so the word 'scientist' was born. This act in itself shows that the social, historical and political context cannot be separated from science. We can mention Einstein and Fermi escaping Germany and Italy due to the racial laws of their respective fascist governments. Emmy Noether, probably the greatest mathematical mind of the twentieth century, was forced to teach under a male colleague's name because she was not allowed to as a woman. How many minds that might have contributed to our shared knowledge never had a chance to get there, and still don't have the chance to do so? Stephen Jay Gould is right when he said, 'I am, somehow, less interested in the weight and convolutions of Einstein's brain than in the near certainty that people of equal talent have lived and died in cotton fields and sweatshops.'

Given that society has decided that my 'personal is political', I have decided to embrace it to the maximum. If this is the only book I write, I wanted a choir of queer voices to accompany me, to break down the complex concepts, to explain what we know of the currently unexplained. This is my way of shining a little light on these other invisible rainbows. The parallels between astronomy beyond the visible and LGBTQIA+ people kept popping up in the making of this book. From the rejection of infrared astronomy by visible-light astronomers in the sixties for not being the right way to do astronomy, to Tana Joseph's peculiar X-ray sources, ignored as they don't fit in with scientists' expectations. Almost too on the nose. And that's how we get to the Universe without shadow. We need to admit that science is not a perfect holder of knowledge. We must constantly challenge what we do know. Only then can we get to the things that we do not know.

The Invisible Is the End of the Beginning

The only parallel I am keen to have you consider is this. This book has made visible so many phenomena, objects and events that exist beyond our senses. And it has also made visible a few dozen brilliant astronomers and astrophysicists from across the world who are also LGBTQIA+. But visibility is not enough for either. For queer people, that means meaningful changes and improvements to society, changes and improvements that would benefit *all* of humanity if implemented. For those scientific subjects, light observations have been amazing, but they have their limits, as we have seen. Yet human ingenuity continues to go beyond not just the limits of our senses but even the limits imposed by the Universe. Light is blocked and absorbed. Lost and scattered. But there is a Universe where no shadow is cast. Welcome to the age of multi-messenger astronomy.

Remember neutrinos? They are a type of fundamental particle that has no electric charge. They are so light that for a long time they were thought to have no mass at all. They interact very rarely and 100 trillions of neutrinos are currently streaming through you every second of every day. They are produced in nuclear reactions from the decay of atoms to the fusion that powers stars, and in extreme events like supernovae. Gravitational waves are created by the motion of every object with mass. They are tiny vibrations in space-time, although the more massive and faster something moves, the more waves it creates. To measure neutrinos, we use enormous vats of pure water and sensitive cameras that can spot a flash when a neutrino bullseye on an atom. One observatory uses the ice in Antarctica as a detector; it is called IceCube – a fantastic name! Cameras are buried within a volume of a cubic kilometre to look for serendipitous flashes of light. IceCube spotted neutrinos coming from extreme celestial events and tracked them to their source, which was then observed in electromagnetic radiation.[1] Gravitational waves are measured with the pulsars and with the detectors we briefly encountered in the radio waves chapter; they are so sensitive to the motion of space-time that waves in the sea tens of kilometres away are a problem. The collision between neutron stars forming the kilonova[2] in the gamma rays chapter was first spotted in gravitational waves and then followed with all the portions of the spectrum of light. Neutrinos, gravitational waves and light are the messengers of this multi-messenger astronomy. We have yet to find an event where all three are used together, but trust me when I tell you it could be any day.

Maybe it has happened between me writing these words and you reading them. Neutrinos and gravitational waves cast no shadows. They do not get stopped as light does. They will allow us to study phenomena like supernovae in a whole new way. We could maybe use them to push beyond the limits of the cosmic microwave background and observe the Universe when it was even younger. Who knows what answers lie there?

In this book we have gained the power to see beyond our limits, and while we discover the next frontier we are also being told that there is a way to access it. Shall we behave in the same manner as our predecessors and exclude these new tools? Shall we only accept that astronomy is done in a certain way and by certain people and that all the rest is not astronomy? I don't think so.

It is a hard, constant process to shed our biases and challenge ourselves. Sometimes it is very annoying. What we know gives us assuredness, a framework, and often comfort. The unknown can be scary. But I embrace the word that Dante gives Ulysses in *The Divine Comedy*: we were not made to live like beasts. We are meant to search for virtue and knowledge. Sure, our ape brain developed to look out for lions in the savannah, so it is very quick to tell us that the unknown is dangerous and that we must protect ourselves against it. But we are also the children of those apes that through bravery shushed that voice and pushed against it. Sure, the unknown can be scary but it can also be wonderful. It can make us better, it can make us grow, and it can make us change. Change is the fundamental state of the Universe and if we stubbornly push against it, we just limit ourselves. I don't like limits. They make me itchy. If there is a horizon, there ought to be something beyond. I desperately want to see what's there, and I hope you do, too.

Notes

RADIO WAVES: The Name of the Game Is Versatility

1 Dr Alfredo Carpineti, 'This Is The First-Ever Image of the Event Horizon of a Black Hole', *IFLScience*, 10 April 2019.
2 South African Radio Astronomy Observatory, 'MeerKAT radio telescope inaugurated in South Africa – reveals clearest view yet of center of the Milky Way', *SARAO*, 13 July 2018.
3 J. Delhaize et al., 'Are giant radio galaxies more common than we thought?', *Monthly Notices of the Royal Astronomical Society*, 14 December 2020.
4 Dane Dodds, Trailer for documentary *!Aitsa*, 2023.
5 P. Natarajan et al, 'First Detection of an Over-Massive Black Hole Galaxy UHZ1: Evidence for Heavy Black Hole Seed Formation from Direct Collapse', *The Astrophysical Journal Letters*, 29 December 2023.
6 Dr Alfredo Carpineti, 'The "Mountains" On A Neutron Star Are Only Millimetres Tall, New Models Suggest', *IFLScience*, 19 July 2021.
7 M.E. Caplan et al. 'Elasticity of Nuclear Pasta', *Physical Review Letters*, 24 September 2018.
8 Dr Alfredo Carpineti, 'First Significant Evidence Of Gravitational Wave Background Using Galaxy-Sized Detector', *IFLScience*, 28 June 2023.
9 D. J. Champion et al., 'Measuring the mass of solar system planets using pulsar timing', *The Astrophysical Journal Letters*, 23 August 2010.
10 Dr Alfredo Carpineti, 'Dark, Starless Galaxy Discovered Accidentally By Astronomers', *IFLScience*, 11 January 2024.
11 J. Margot et al., 'Spin state and moment of inertia of Venus', *Nature Astronomy*, 29 April 2021.

12 Dr Alfredo Carpineti, 'Tiny Asteroid To Hit Earth Over Siberia Today – Here's What To Know', *IFLScience*, 3 December 2024.

13 James Felton, 'Is Bennu Really Going To Hit Earth On September 24, 2182?', *IFLScience*, 18 September 2023.

14 Jonah S. Peter et al., 'Detection of HCN and diverse redox chemistry in the plume of Enceladus', *Nature*, 12 January 2023.

15 Joseph R. Michalski, 'Diverse volcanism and crustal recycling on early Mars', *Nature Astronomy*, 12 February 2024.

16 A. O. Warren and E. S. Kite, 'Narrow range of early habitable Venus scenarios permitted by modeling of oxygen loss and radiogenic argon degassing', *Proceedings of the National Academy of Science*, 6 March 2023.

17 Dr Alfredo Carpineti, 'Moderately Advanced Aliens Could Easily Take Over A Milky Way-Like Galaxy, Simulations Show', *IFLScience*, 17 June 2021.

18 Dr Alfredo Carpineti, 'We Might Not See The Solar Panels Or Dyson Spheres Of An Alien Civilization', *IFLScience*, 5 August 2024.

19 Dr Alfredo Carpineti, 'Earth Just Received Laser-Beamed Cat Video From 19 Million Miles Away', *IFLScience*, 19 December 2023.

20 Kelly Kowalski, 'Are Extraterrestrials Listening In On Our Phone Conversations?', *IFLScience*, 8 September 2023.

21 C. D. Trembley, 'COSMIC: An Ethernet-based Commensal, Multimode Digital Backend on the Karl G. Jansky Very Large Array for the Search for Extraterrestrial Intelligence', *Astronomical Journal*, 27 December 2023.

22 Rebecca McDonald, 'Unlocking the Secrets of Fast Radio Bursts: Scientists Find More Pieces to the Puzzle of Mysterious Space Signals', *SETI*, 13 December 2023.

23 E. Petroff et al., 'Identifying the source of perytons at the Parkes radio telescope', *Monthly Notices of the Royal Astronomical Society*, 29 June 2015.

MICROWAVES: The Bigger Picture

1 John G. Gramer, 'Neutrinos and WIMPs', *Analog Science Fiction and Fact*, 20 October 1985.
2 Swinburne University of Technology, Cosmic Microwave Background Dipole | COSMOS', *The SAO Encyclopedia of Astronomy*: https://astronomy.swin.edu.au/cosmos/*/Cosmic+Microwave+Background+Dipole
3 Adam G. Riess et al., 'JWST Observations Reject Unrecognized Crowding of Cepheid Photometry as an Explanation for the Hubble Tension at 8σ Confidence', *The Astrophysical Journal Letters*, 6 February 2024.
4 Dr Alfredo Carpineti, 'Hubble Tension Solved? Astronomers Race To Save Standard Model Of Cosmology', *IFLScience*, 25 September 2024.
5 Dr Alfredo Carpineti, 'Are We Really All Made Of Stardust?', *IFLScience*, 30 September 2023.
6 D. W. Pesce et al., 'The Megamaser Cosmology Project. XIII. Combined Hubble Constant Constraints', *The Astrophysical Journal Letters*, 26 February 2020.
7 R. Teague et al., 'Meridional flows in the disk around a young star', *Nature*, 16 October 2019.
8 O. Balsalobre-Ruza et al., 'Tentative co-orbital submillimeter emission within the Lagrangian region L5 of the protoplanet PDS 70 b', *Astronomy & Astrophysics*, 6 June 2023.

INFRARED: The Universe at the Edge of Our Senses

1 B. Welch et al., 'A highly magnified star at redshift 6.2', *Nature*, 30 March 2022.
2 Marc Airhart, 'JWST Reveals Milky Way-like Galaxies in Young

Universe', *UT News*, University of Texas at Austin, 5 January 2023.
3 Dr Alfredo Carpineti, 'JWST Shows Gas Around A Supermassive Black Hole Like Never Before', *IFLScience*, 22 July 2022.
4 Dr Alfredo Carpineti, 'New Record-Breaker: Earliest And Most Distant Galaxies Ever Found Spotted By JWST', *IFLScience*, 3 June 2024.
5 Judith Adkins, '"These People Are Frightened to Death" – Congressional Investigations and the Lavander Scare', National Archives, Summer 2016.
6 C. H. Humelsine, 'Memorandum from Carlisle H. Humelsine to James Webb', *National Archives Catalog*, produced 24 June 1950.
7 Dr Chanda Prescod-Weinstein, Dr Sarah Tuttle, Dr Lucianne Walkowicz and Dr Brian Nord, 'The James Webb Space Telescope Needs to Be Renamed', *Scientific American*, 1 March 2021.
8 Phil Plait, Twitter Thread on Telescope Names, 7 April 2020.
9 H. Holden Thorpe, 'Hard, not easy', *Science*, 15 December 2022.
10 Matthew Francis, 'The Problem With Naming Observatories For Bigots', *Forbes*, 11 June 2015
11 Dr Jason Wright, 'Why NASA should have a do-over on the name of JWST', *AstroWright*, 25 October 2022.
12 Dr Alfredo Carpineti, 'NASA To Stick With Controversial Name For New Space Telescope', *IFLScience*, 30 September 2023.
13 Brian C. Odom, 'NASA Historical Investigation into James E. Webb's Relationship to the Lavender Scare', 18 November 2022.
14 Alexendra Witze, 'Exclusive: Documents reveal NASA's internal struggles over renaming Webb telescope', *Nature*, 25 March 2022 and full documents: https://media.nature.com/original/magazine-assets/d41586-022-00845-6/20247466
15 E. A. Cech and T. J. Waidzunas, 'Systemic inequalities for LGBTQ professionals in STEM', *Science Advances*, 15 January 2021.
16 American Physics Society, 'APS LGBT Climate in Physics Report', March 2016.

17 Bethany Downer, 'Hubble finds that a black hole beam promotes stellar eruptions', *ESA Hubble*, 6 September 2024.
18 Dr. Markus Pössel, 'Milestone for JWST exoplanet observations: atmosphere properties in more detail than ever before', *Max Planck Institute for Astronomy*, 22 November 2022.
19 Dr Alfredo Carpineti, 'Hubble Tension Solved? Astronomers Race To Save Standard Model Of Cosmology', *IFLScience*, 25 September 2024.
20 Dr Alfredo Carpineti, 'Challenge To Theory Of The Universe Reignited In New Publication', *IFLScience*, 10 December 2024.
21 'The Euclid Mission', European Space Agency.
22 'Origins Space Telescope – Mission Concept Study Report', NASA, August 2019.

ULTRAVIOLET: The Blinding Limelight

1 Dr Alfredo Carpineti, 'Exclusive: Astronaut Ed Gibson On How Skylab, The First US Space Station, Changed Space Exploration', *IFLScience*, 25 May 2021.
2 L. Lamy, 'Auroral emissions from Uranus and Neptune', *Philosophical Transactions of the Royal Society A: Mathematical, Physical and Engineering Sciences*, 9 November 2020.
3 H. Melin et al., 'Discovery of H_3^+ and infrared aurorae at Neptune with JWST', *Nature Astronomy*, 26 March 2025.
4 J. T. Trauger, 'Saturn's Ultraviolet Aurora', *Hubble Site*, 7 January 1998.
5 L. Pallier and R. Prangé, 'More about the structure of the high latitude Jovian aurorae', *Planetary and Space Science*, August 2001.
6 Z. Girizian et al., 'Discrete Aurora at Mars: Dependence on Upstream Solar Wind Conditions', *Journal of Geophysical Research: Space Physics*, 27 March 2022.
7 Marina Galand, 'Unique ultraviolet aurora spied at Rosetta's comet', *European Space Agency*, 21 September 2020.
8 Dr Alfredo Carpineti, 'This Is The Best Ever Image Of The

Heaviest Star In The Known Universe', *IFLScience*, 25 May 2021.
9 Roxana Bardan, 'New NASA Mission will Study Ultraviolet Sky, Stars, Stellar Explosions', *NASA*, 13 February 2024.

X-RAYS: More Things in Heaven and Earth

1 Katy Evans, 'Parker Solar Probe Phones Home! Spacecraft Survives Record-Breaking Closest Pass To The Sun', *IFLScience*, 27 December 2024.
2 D. Lim et al., 'The Role of High-frequency Transverse Oscillations in Coronal Heating', *The Astrophysical Journal Letters*, 24 July 2023.
3 'How Big is the Magnetosphere? – Mission Juno', *NASA*.
4 G. D. Reeves et al. 'Energy-dependent dynamics of keV to MeV electrons in the inner zone, outer zone, and slot regions', *JGR Space Physics*, 28 December 2015.
5 S. Aizawa et al., 'Direct evidence of substorm-related impulsive injections of electrons at Mercury', *Nature Communications*, 18 July 2023.
6 A. Merloni et al., 2024, 'The SRG/eROSITA all-sky survey: First X-ray catalogues and data release of the Western Galactic hemisphere', *Astronomy & Astrophysics*, February 2024.
7 Dr Alfredo Carpineti, 'Star With The Shortest Orbital Period Around Supermassive Black Hole Discovered', *IFLScience*, 6 July 2022.
8 M. C. Begelman, J. R. King and J. E. Pringle, 'The Nature of SS433 and the Ultraluminous X-ray Sources', *Monthly Notices of the Royal Astronomical Society*, 17 July 2006.

GAMMA RAYS: The Pursuit of the Energy of the Extreme

1 M. Piersanti et al., 'Evidence of an upper ionospheric electric field perturbation correlated with a gamma ray burst, *Nature Communications*, 14 November 2023.
2 Dr Alfredo Carpineti, 'A Supernova 50 Light-Years Away Could Start A Mass Extinction', *IFLScience*, 15 May 2017.
3 H. Svensmark, 'A persistent influence of supernovae on biodiversity over the Phanerozoic', *Ecology and Evolution*, 16 March 2023.
4 'Gamma-ray bursts: are we safe?!', *European Space Agency*, 17 September 2003.
5 P. Padovani, 'Active Galactic Nuclei at All Wavelengths and from All Angles', *Frontiers in Astronomy and Space Sciences*, 7 November 2017.
6 G. Madejski and M. Sikora, 'Gamma-Ray Observations of Active Galactic Nuclei', *Annual Review of Astronomy and Astrophysics*, 10 August 2016.

EPILOGUE: Into a Universe Without Shadows

1 Jonathan O'Callaghan, 'Incredible "Ghost Particle" Discovery Heralds A New Era In Astronomy', *IFLScience*, 12 July 2018.
2 Dr Alfredo Carpineti, 'In A Historic Discovery, A Neutron Star Merger Has Been Observed For The Very First Time', *IFLScience*, 16 October 2017.

Acknowledgements

Thanks to my sister Francesca, whom I could trust to tell me the truth about my writing and convince me that I had something here by saying that the radio chapter 'was actually pretty good'. We shall not comment on the first version of the X-ray chapter.

A mamma e papà per tutto il loro amore e il loro sostegno, non solo per il libro ma da sempre!

To my dearest friends and beta readers, Russell and Katy, who provided diametrically opposite feedback on what was needed to improve the book. Somehow, I believe I was able to do a Hegelian synthesis of the two! (Not happy you made me mention Hegel.)

To Sacha, who gave me the right advice and put me on a turbulent, but eventually correct, path! You were instrumental in this happening!

To Aliya, who commissioned this book for Unbound and was absolutely the best, you are a star!

Thank you to the Wilton Square Books team, Will, Anna, Dan and Dan for getting this over the finish line. Thanks to my editor, Richard, for translating Alfredoish into English, and thanks to my agent, Sam, for the support and the excitement that came with new plans.

To Simon, for the virtual hobby nights where I wrote and you painted, and to Craig for the many phone calls commiserating about the things that were not working. To Xenia, Robin, Davide B, Francesco, Davide C, Sara, Carl, Brad, Steve, Dean, Gabe, Chris K and Henry for the staunch support throughout these years. To all friends, family, and colleagues who listened to the rants about this book. It has not been an easy ride, but thank you for sticking around.

To all the brilliant people who leant me their time to be interviewed: Affelia Wibisono, Alicia Roucco-Escorial, Aman Chokshi, Bradley Meyers, Dan Wilkins, David Williams, Ed Rivera-Valentin, Franck Marchis, Jake Taylor, Kaew Samaporn Tinyanont, Keir Birchall, Madalyn Johnson, Marie Wingyee Lau, Max Silva-Feaver, Mike Gladders, Murray Brightman, Peter Coles, Rahul Jayaraman, Rami Mandow, Sally Oey, Scott Gaudi, Sthabile Kolwa, Tana Joseph, Tanmoy Laskar, Taylor Hutchison, Wladymir Lyra, Yang Chentao, Yaswant Devarakonda, Yuzhe Song. An extra special thank you to Jake for actually reading some of the book and providing feedback. May this rainbow be visible for all to see.

To Llywelyn Orpheus 'Clue' Carpineti for demanding distracting, very important walks!

And lastly, my husband Chris, for his unwavering support over these three years of book process, as well as reading the draftiest drafts, and dealing with all the ups and all the downs. He would have been first, but he said that in all the books he reads, the partner is mentioned last, so here he is… maybe those authors don't have quite as good a partner as I do!

The Book Was Possible Thanks To:

Patron: Dan McNeil

Supporters of the book:

Adam McGreggor
Alan Sunley
Alasdair Barclay
Andrea Bandelli
Andrew E. Bryan
Andrew Hall
Andrew Moore
Anne Jones
Anthony Rimmer
Avery Cunningham
Bilaal Mohammed
Brendan Murphy
Caleb Walkley
Chris Barrett
Clara Feutrier
Conor McKeever
Dan McNeil
Dan Murray
Dana Love
David Black
David Pegg
David Roskin
Dean Newby
Duncan
Ed Sum (he/him)
Emily Harford
Eugene Lynch
Fabian Albrecht
Francesco & Federica
Frank Coates
Gary Hicks
Gavin Wilson
Gayle Yeomans
George Shaw
Gill England
Gregory Kroll
Helen Umberger
Ian Betteridge
James Gregory-Monk
Jamie Barrett
JC Candanedo
Jen Denison
Jennifer Milward
João Teixeira Duarte
John Cooke
Jon Biggs
Joshua Gladwin
Katy Evans
Kevin Kane
Kostas Glaros
Laura Rüegger
Laura Tobin
Lee Willis
Lucas Fothergill
Luke
Luke Fuller
Mark Vent
Marzi Pečen
Michael Hewitt & Jean Vianney Cordeiro
Michael J. Wilson
Michael Trueman
Michelle Hughes
Mike Williams
Neil McKenna
Nicole Burstein
Nicholas Fahey
Nick Proellochs

Omar Rafael Regalado Fernández
Pam Kingsley
Paul Ian Cross
Pete Shorney
Polis Loizou
Rachael
Rich Hobson
Roberto Efraín Díaz
Rory Queripel
Ross Brooks
Ryan Unruh
Simon Ashe
Simon D. A. Clark
Stephen C
Stephen Usins Yeardley
Steven Di Costa
Stuart Whitehouse
Susanne Emde
Thomas Ganter
TJ Ronningen
Tomas Persson
Vanessa Heggie
Xenia De Lucia

Index

A Note on the Author

Dr Alfredo Carpineti (he/him) is a queer Italian astrophysicist, science journalist, and social activist. He has an M.Sc. in Quantum Fields and Fundamental Forces and a Ph.D. in Astrophysics from Imperial College London.

He is the Space & Physics Editor for *IFL Science*, where he has worked for ten years, writing over 6,000 articles that have been read by millions. He is the chair and founder of Pride in STEM, an award-nominated British charity dedicated to supporting and showcasing LGBTQIA+ in STEM.

A Note on the Type

The main text of this book is typeset in Adobe Caslon Pro. William Caslon (1692–1766) was an English typefounder who released his first typefaces in 1722. These types were based on seventeenth-century Dutch old-style designs, used extensively in England, at the time. Caslon's serif types became popular throughout Europe and the American colonies for their quality and their attractive appearance, suitable for extended passages; the first printings of the American Declaration of Independence and the Constitution were set in Caslon.